Imaging Hoover Dam

Imaging Hoover Dam

The Making of a Cultural Icon

Anthony F. Arrigo

UNIVERSITY OF NEVADA PRESS

Reno & Las Vegas

University of Nevada Press, Reno, Nevada 89557 USA
Copyright © 2014 by University of Nevada Press
All rights reserved
Manufactured in the United States of America
Design by Kathleen Szawiola

Library of Congress Cataloging-in-Publication Data

Arrigo, Anthony F., 1972-
 Imaging Hoover Dam : The Making of a Cultural Icon / Anthony F. Arrigo.
 pages cm
 Includes bibliographical references and index.
 ISBN 978-0-87417-953-8 (cloth : alk. paper) — ISBN 978-0-87417-954-5 (e-book)
 1. Hoover Dam (Ariz. and Nev.) — In art. 2. Hoover Dam (Ariz. and Nev.) — In popular
 culture. I. Title.
 N8214.5.U6A72 2014
 704.9'4479313 — dc23   2014010700

The paper used in this book meets the requirements of American National Standard for
Information Sciences — Permanence of Paper for Printed Library Materials, ANSI/NISO
z39.48-1992 (R2002). Binding materials were selected for strength and durability.

FIRST PRINTING

23   22   21   20   19   18   17   16   15   14
5   4   3   2   1

For my beloved girls: Sarah, Bonnie, and Clara

Contents

List of Illustrations

Preface

A few years ago I was having a casual conversation with someone about a recent vacation to Nevada and California in which I took a side trip to see Hoover Dam. Little did I know that the seemingly inconsequential exchange—a polite "how was your trip" type of encounter that we have with people all the time—would become the catalyst for several years of research culminating in this book.

My interlocutor, not being from the United States and upon hearing that I'd stopped at Hoover Dam, said, "Oh, you must be very proud of that, as an American." This curious response prompted a longer discussion in which we surmised that it probably wouldn't be a stretch to say that, as with Niagara Falls or Mount Rushmore, virtually every American can recognize Hoover Dam and probably knows at least something about it. In fact, it might be the only dam they know by name. The interesting question to me was why.

Intrigued to learn more about the dam, what I quickly found was that, although the history was interesting, what really fascinated me were the images: the riveting depictions of laborers scaling thousand-foot canyon walls to set their blasting caps, the celebratory tone of images showcasing the dam's construction technologies, the pitiable photographs of families living in squalor around the work site. I was struck by the rhetorical features of these images, as well as their ability to instantaneously arrest my attention and "speak" to me in ways that the text did not or could not. As I began to research the topic in more depth, I found a seemingly endless trove—thousands upon thousands—of images of all sorts published in every conceivable medium. It seemed to me that these images had their own story to tell, and the more images I found, the more diverse the story became, in some cases veering wildly from the received history of the dam that I had casually come to know.

One thing that struck me as curious, however, was that although Hoover Dam is a nearly universally known—indeed iconic—structure, I couldn't identify any particular image as *the* iconic image of the dam. Furthermore, although a few scholars had tackled individual slices of Hoover Dam imagery, no one had attempted a comprehensive analysis of its whole visual repertoire. Consequently, by identifying some key themes that emerge from the vast sweep of Hoover Dam imagery, this work is a step toward filling that gap.

In this book I try to answer the question of how Hoover Dam evolved from a pipe dream of land developers and farmers, to an ambitious civil engineering project in the middle of the Mojave Desert, to the visual and cultural icon that it is today. To do this, in contrast to most scholarship on the dam, I provide a significant shift in focus away from chronicled accounts of how it was built and onto its myriad visual representations. In doing so, I trace how its imagery was deployed through advertising, government propaganda, journalism, and other promotional outlets to shape the public's perception of the project. This discussion ranges from how the dam's imagery reflects the cultural and ecological imperatives that precipitated its construction, to the influence of religious doctrine and the American agrarian movement in the drive to build the dam, to the visual commodification of the project as a way to sell cars, trucks, vacations, and a variety of other goods and services. Ultimately I find that approaching the dam's construction through its imagery juxtaposes its traditional narrative of economic and industrial triumph with an often-contrasting visual counter-narrative of technological domination of both humans and the natural environment.

In order to identify the arguments advanced in the visual-verbal discourses surrounding the dam, this book draws on a variety of primary sources gathered over a several-year period at a number of archives and repositories, including the County of Los Angeles Public Library, the California State Library, the Boulder City Museum and Historical Association, the Nevada State Museum, the Herbert Hoover Presidential Library and Museum, the US National Archives, the University of Nevada at Las Vegas (UNLV), the University of Nevada at Reno, Archives of the Los Angeles Department of Water and Power, the University of Southern California Libraries Special Collections, the Wilson Library at the University of Minnesota, and the Bancroft Library at the University of California at Berkeley.

I would like to acknowledge several people in particular for their help in this endeavor, including Katherine Adams at the County of Los Angeles Public Library; Catherine Hanson at the California History Room, California State Library; Colleen Dwyer, public affairs specialist, US Bureau of Reclamation; David Kessler at the Bancroft Library, University of California at Berkeley; Wendy Stevenson, reference librarian at the State Historical Society of Iowa; Carol Kirsch, supervisor of libraries, special collections, and publications at the State Historical Society of Iowa; and Dace Taube, assistant head of special collections at the Doheny Memorial Library at the University of Southern California. I am indebted to Bernadette Longo, Daniel Philippon, and Larry Baker for their feedback, suggestions, patience, and encouragement. I am also grateful to Beth Ayer for her work in helping me with permissions for many of the images found in this book. Special thanks are in order for Dennis McBride, director of the Nevada State Museum, Las Vegas, who helped me immensely in finding information on Winthrop A. Davis and for his quick and insightful e-mail responses to inquiries on a variety of other topics. Also deserving particular thanks is Laura Hutton, museum coordinator at the Boulder City Museum and Historical Association, who scanned numerous images that you see in this book, and who helped me navigate the wonderful archival materials they have available in Boulder City; this book simply would not have been possible if not for Laura's and the Boulder City Museum's generosity of time and access to their collections. Thanks also go to Delores Brownlee, public services and photographs assistant, at the University of Nevada, Las Vegas, Special Collections, who graciously coordinated the permissions and reproductions for all of the images you see in this book from the UNLV archives.

I would also like to thank Matt Becker at the University of Nevada Press for his guidance in everything from the publishing process to how to be a professional writer. I would also like to especially thank Richard Graff, who dedicated many, many hours to the initial phase of this project and whose editorial and thematic comments, discerning suggestions, and commitment to excellence made this book far better than I could have ever hoped to accomplish on my own.

Finally, and most important, I would like to acknowledge my loving wife, Sarah, my two wonderful daughters, Bonnie and Clara, and my entire family for their support, interest, and encouragement in this process. I truly could not have finished this book without you.

Imaging Hoover Dam

Introduction

Beginning at least a decade prior to the start of construction, through the period of its creation, and continuing even up to today, Hoover Dam has been repeatedly, relentlessly "imaged." This book tracks how the images of Hoover Dam have depicted and documented its existence from conception to sublime physical reality to its prominent place in the American consciousness as an icon of engineering achievement. Over the course of several decades, thousands of representations of Hoover (and at various points in its history also referred to as Boulder) Dam have formed a wide-ranging tableau that spans the entirety of its construction, one that is set firmly within the pantheon of American visual iconography. This book seeks to explain how the dam reached such a status by considering its imagery as rhetorical, or, in other words, how governments, businesses, and journalists used images of the dam to shape the perceptions and attitudes of Americans regarding technology, reclamation projects, tourism, engineers, laborers, minorities, nature, and the construction of the dam itself.

Over the years, images of the dam have served several rhetorical purposes. Some of these purposes were quite explicit, others rather subtle. Many images worked in consort with verbal arguments for the immediate effect of gaining public support for the project, promoting tourism, or advertising consumer goods. At the same time, however, these images attended to broader and more profound social effects that speak to how images can work to frame the public's cultural memory, structure the hierarchies of citizenry, and symbolize or crystallize ideologies of the state. Moreover, many

Figure 0.01. "Black Canyon—Hoover Damsite: Looking upstream thru Black Canyon from a point of rock on NV side one-half mile below lower Hoover damsite. Lower damsite is at prominent point on right," April 12, 1922 (photograph, Walker R. Young, Bureau of Reclamation). Courtesy of the Boulder City Museum and Historical Association.

of the images were indeed visualizations that served to reify values, beliefs, and aspirations shared by a community, region, and in some cases the entire nation.

Even before its completion, there was no doubt that Hoover Dam was destined for renown. In the years and decades before the first blasts of dynamite began to reshape the walls of Black Canyon (figure 0.01) in preparation for what was to become the largest dam in the world, it was proclaimed a savior of the Southwest, an unrivaled feat of engineering, and, according to the *Los Angeles Times*, a "gateway of Empire." It was repeatedly hailed as a world wonder, indeed a sublime paean to American technological progress and a symbol of American ingenuity and pride. Senator Hiram W. Johnson, coauthor of the Swing-Johnson appropriations bill that legislated the dam's construction, wrote in 1928 that once completed it would be "the greatest constructive project of our generation. There is nothing comparable to it within our memories, save the construction of the Panama Canal. It is a project of national importance."[1] The *Los Angeles Times* took an even more expansive perspective, proclaiming in October of 1933, "This great structure presents a picture of massive power, which overwhelms even the modern concept of the great Mayan builders." Surpassing the Great Wall of China, the Acropolis, Hagia Sophia of Constantinople, and the pyramids of Egypt, the *Times* declared the dam to be "in fact, the greatest structure ever built by man."

Nevertheless, Hoover Dam's construction was not a foregone conclusion. Many questioned the legality of the Bureau of Reclamation undertaking such a project while complaining of a (real or perceived) federal disregard for states' rights. Some doubted that the federal government would be able or willing to pay the huge sums of money that would be required to construct the dam in the midst of the worst economic depression in US history. Others questioned whether the economic windfall predictions from the project would actually come to fruition. Some doubted that it was even possible to build such a structure because the engineering challenge was simply too great, while still others had profound suspicions about the growing clout of Los Angeles and its potential to take an inequitable share of electricity and water once the project was completed. To counter these perceptions, governments, news media, and business interests across the Southwest embarked on a campaign to sway public opinion by deploying visual and verbal narratives projecting romantic visions of what might be, or what could or would eventually become reality, a campaign that would include as its centerpiece a multiphase arc of Hoover Dam imagery that would unfold over the course of the next several decades.

Years before construction even began, creative renderings, often based on preliminary engineering schematics and grainy photographs of the proposed canyon site (also frequently coupled with promissory verbal descriptions), expressed sanguine projections of bountiful crops, limitless water and electricity, booming businesses, and extravagant luxuries resulting from the dam. Once it was approved and construction was underway, images of the dam shifted to primarily documentary-style photographs that continued an intense publicity campaign by the Bureau of Reclamation to celebrate the heroic efforts of its engineers and glorify their technological accomplishments and innovations. Likewise, by using the dam as a visual marker to imbue products with the same emotional and cultural attachments that people had to the structure itself, Hoover Dam imagery also found its way into advertisements for an array of products as businesses sought to capitalize on America's pride and acknowledgment of the project as one of its greatest achievements. Finally, after its completion, the Department of the Interior sought to pivot toward another narrative of the dam and its man-made reservoir, Lake Mead, as an outdoor recreation and vacation destination—America's "playground" in the Southwest.

Through all of its stages of development, Hoover Dam was sketched, painted, sculpted, photographed, rendered, and appropriated countless times. Beginning in the 1920s and continuing for decades thereafter, an enormous number of Hoover Dam–related images were used in virtually every visual medium. Branches of the federal government, including the Bureau of Reclamation and the National Parks Service, deployed professional photographers to chronicle the dam's construction for technical reasons, but also for promotional—some would say propagandistic—purposes, in order to sell the project to elected officials and the American public. Proponents of the project (both local and national) carefully selected and placed particular images in newspapers, pamphlets, booklets, and texts of all sorts to promote everything from electric utilities, to vacationing and travel, to the New Deal. The Bureau of Reclamation also produced hundreds of images of the dam for on-site visitors that ranged from entrance permits and promotional decals to maps, buttons, booklets, brochures, and an array of souvenirs. The US Postal Service also placed images of the dam on two of their stamps, thus ensuring that tens of millions of impressions of the dam's image circulated around the country and beyond.[2]

Some of the most well known American artists of the early twentieth century also produced works portraying Hoover Dam. In his painting *Conversation: Sky and Earth* (1940), Charles Sheeler, one of the pioneers of American modernism, depicts electrical transmission towers against the backdrop of the dam as part of a series for *Fortune* magazine celebrating American industrial productivity. Stanley Wood produced a series of paintings under the sponsorship of the Public Works Art Project that were also published in *Fortune* for a separate pictorial essay specifically on the Hoover Dam project. William Gropper's ten-by-thirty-foot mural *The Construction of a Dam* installed in 1939 in the Department of the Interior building in Washington,

Figure 0.02. "Photograph of the Boulder Dam from Across the Colorado River," 1941 (photograph, Ansel Adams, Department of the Interior, National Park Service). Courtesy of US National Archives, Branch of Still and Motion Pictures.

DC, which pays homage to both the contributions of labor and the importance of industry, is based in part on his sketches made during a trip to the dam in 1937.[3] Hugh Ferriss, the creator of highly influential modernist depictions of skyscrapers and other city scenes, many of which appeared in his now-famous book *Metropolis of Tomorrow*, sketched several renderings of Hoover Dam as part of a 1941 grant from the Architectural League of New York to chronicle works in contemporary architecture. Ansel Adams also took photographs of Hoover Dam in 1941 for the US National Park service, one of which, "Photograph of the Boulder Dam from Across the Colorado River," has become one of the more celebrated images of the dam (figure 0.02).

Although representations of Hoover Dam appeared in texts of all kinds, newspapers were the primary medium in which they were placed for mass public viewing. They were also the main venue in which public battles over the dam's construction played out. In addition to untold news stories and opinion pieces about the project, thousands of sketches, renderings, and photographs of the dam and its construction appeared in hundreds of newspapers across the country. Images of the dam also appeared in magazines ranging from industry-specific trade publications, such as *American Steel & Wire*, *Engineering News Record*, and *Western Construction News and Highways Builder*, to widely circulated periodicals aimed at the general public, such as *Popular Mechanics*, *Fortune*, *Time*, *Life*, and *Harpers*, plus travel magazines of all sorts.

Advertisers also used depictions of the dam in numerous promotions. In addition to the scores of Hoover Dam images deployed by all levels of government for their own publicity purposes, companies such as Alcoa, Westinghouse Electric, TWA, Pan Am, Pontiac, Studebaker, R. J. Reynolds Tobacco Co., and many others used images of the dam as part of their advertising and promotional campaigns, some long after the dam itself was completed. For example, in the 1960s, Sunoco featured the dam on one of its twenty "Landmarks of America" coins for patrons of their gasoline stations, and Coca Cola used Hoover Dam images on the inside of bottle caps as part of their "Travel the USA" series, in which people were encouraged to collect all 102 different destinations. In 1985, Coke also produced a series of items emblazoned with images of the dam commemorating its golden anniversary, including a fiftieth-anniversary metal serving tray and special Coca Cola bottles housed in commemorative Hoover Dam boxes. The Topps company, best known for its sports-related cards and memorabilia, issued an "Era Icons" set that includes a colorful, art deco–style illustrated interpretation of Ansel Adams's 1941 photograph, along with the official US Postal Service three-cent Hoover Dam stamp embedded into the card. Topps also issued a "Hoover Dam Construction Completed" trading card with an image of the dam at the top and an "authentic buffalo nickel" encased in a plastic covering at the bottom.

Although many of its images are considered high art, with photographs and paintings of the dam and its construction appearing in museums and art collections around the world, Hoover Dam is also well represented in the category of what can only be described as kitsch. Innumerable trinkets with representations of Hoover Dam on them have surfaced over the years, ranging from the typical postcards, beer and coffee mugs, shot glasses, T-shirts, hats, coasters, and key chains to more rarified items such as book ends, "Fine

Staffordshire Ware" sets of china (and numerous other varieties of collector's edition plates), pearl-handled pocket knives with an engraved image of the dam, and keepsakes of all sorts, including Christmas tree ornaments, desktop pen holders, bell chimes, and toothpick holders. Union Pacific Railroad produced commemorative stainless steel Hoover Dam flasks and cigarette-carrying cases. In 1969, Jim Beam issued a special four-fifths-quart-sized Kentucky straight bourbon whiskey Hoover Dam decanter. There are Hoover Dam snow globes, souvenir travel patches, bronze cast Hoover Dam ashtrays, felt college-style pennants, miniature Dutch Hoover Dam wooden shoes, Hoover Dam leather-pouched sewing kits, belt buckles, salt and pepper shakers, car thermometers, and much more. One of the strangest Hoover Dam–related items appeared in 1983 for the fiftieth anniversary of the first batch of concrete poured at the dam in which Fisher company issued their commemorative Hoover Dam "space pen," the pen "so advanced" that it will write upside down and is "guaranteed to write in the gravity free void" or your money back!

Given this endless array of images, it is no exaggeration to say that Hoover Dam was, and continues to be, hypervisualized. The project's extensive visual record provides us with a rare instance in which thousands of representations of a single undertaking were produced by a variety of interested parties and deployed over a multidecade period, often for persuasive purposes and with ramifications for an entire nation. Although the Hoover Dam project itself has been the subject of hundreds of publications,[4] this mass of related visual material has barely been considered.[5] By foregrounding representations of Hoover Dam that were produced before, during, and after its construction, *Imaging Hoover Dam* shows how this supra-discursive visualizing process was integral to the development of the mythology, indeed the very iconicity, of the dam, and how the use of Hoover Dam imagery shifted over time from ensuring its construction, to its celebration as a sublime engineering wonder, to its utter commoditization as a means of selling everything from whiskey, to cars, to vacations, to space pens.

ONE OF THE PRIMARY CONTRIBUTIONS of this book to the literature on Hoover Dam is that it explores how the visual culture of the project worked to establish the dam as an indelible icon for millions of Americans in a way that verbal descriptions alone simply could not. Central to this effort was the capacity of visual messages to meet the shifting goals of corporate and government entities, as well as to portray an ideology of domination through sublime technology—a mythos that encapsulated long-held American cultural attitudes and religious principles regarding nature, technology, and regional and national identity. Although the idea of a giant dam to finally control the Colorado River's flood-then-drought cycle (a phenomenon that had exasperated settler-farmers in the Southwest for decades) was the topic of presidential speeches, congressional debates, and innumerable newspaper articles and opinion pieces, I argue that it was instead its awe-inspiring imagery that played the decisive role in ensuring the dam's construction and its permanent place as America's crowning engineering achievement of the early twentieth century. This book, then, establishes Hoover Dam's imagery as the central constitutive element in its cultural mythology and considers the absolute necessity of using images instead of, or in congruence with, words to convey the importance and grandeur of the project.

Although advocates of the dam promised to finally tame the "menacing" Colorado River while at the same time providing vast wealth for the Southwest, it was the confluence of media, business, and government-created imagery that propelled its construction and assured its status as an enduring icon of Americanism and modernity.[6] A key claim toward this end is that the dam reached such a status not by any single image, but through an accumulation of images over a significant period of time. Unlike other studies that center on a single photographer, for example, or on the most celebrated and reproduced individual works of photography, this book seeks to show that a cultural icon can gain such a status not just by a specific image, but also through an accumulation of imagery—a "frenzy of seeing" what is real (indexical) and becoming-real—and through visuals of several different types: illustrations, composite renderings of photos/sketches, and so forth, not just photographs.[7]

Additionally, few projects in American history have had such extensive documentation (both textual and visual) as Hoover Dam, yet the dam does not have a single, agreed-upon "iconic" image that dominates its cultural representation. Although Hoover Dam images function similarly to Hariman and Lucaites's conception of iconic photographs with respect to their cultural force, political influence, symbolism, representation, reproduction, identification, and historical narrativity, there is no single iconic photograph of the dam or its construction.[8] Instead, the rhetoric of the dam project develops through its hypervisualization in a range of modalities and through an accumulation of imagery over time. Whereas iconic photographs often start as a singular entity—a seminal photographic snapshot that is then continually refined and reappropriated over time within the public sphere—Hoover Dam's imagery has worked in the opposite direction by beginning with a host of images (illustrations rendered by human hand, photographs, or composites of the two) that changed and morphed diachronically, eventually aggregating into a generalized, socially imagined montage of not only the completed dam but the entire course of construction, a process that unfolded over many years.

Nevertheless, although the dam's imagery did (and does) shape civic identity, it did so initially by promulgating the ideologies of powerful institutions through a rigorously controlled and relentless scopic regime that visualized humans, technology, and nature. These types of images cohered with an institutionally sanctioned, imperialistic agenda of empire in the Southwest set forth by both political and business interests, an agenda that was reified in a variety of visual and verbal forms and circulated widely to the public. Acutely aware that imagery was a potent element of their drive to control the Colorado River (an accomplishment that would provide near-hegemonic power in the Southwest), businesses, government, and news media produced visual and verbal messages that lionized engineers while minimizing workers and venerated a sublime technology while devaluing the landscape, all while overlooking the project's iniquitous labor practices and discriminatory hiring policies in the service of maximizing profits for contractors and fulfilling the government's promise of wealth and bounty for the region.

To understand the full range of the visual imagery of Hoover Dam, it is also important to examine not only what is seen but also what is unseen. Such considerations remind us that the act of imaging, whether through painting, illustration, or photography, always has behind it a human actor, often working under ideological and institutional controls. Here we must heed Roland Barthes's question, "Of all objects in the world: why choose this object, this moment, rather than some other?"[9] This is particularly important when considering the internal and external shaping mechanisms of an image and how they might be working in the service of maintaining institutional power structures, political and cultural ideologies, and certain scopic regimes. For example, it has been well established that governments and businesses were complicit in eschewing laws designed to provide for the safety of laborers in exchange for faster, less costly construction. However, what is largely ignored is how corporate and government interests actively worked—and largely succeeded—in maintaining this hegemony over laborers by rigidly suppressing the distribution of images depicting labor strikes, industrial accidents, questionable construction practices, or anything else that would run counter to the prevailing narrative of safety, efficiency, and a prideful workforce.

Imaging Hoover Dam also advances the position that culture is forcefully entwined with nature and that nature, too, could and should be an important consideration in analyses of culture. Although I discuss at length how the visual-verbal culture of Hoover Dam was constitutive of race, class, gender, and political and social institutions, any discussion of the dam's representations also necessitates considerations of the environment, a position that unfortunately too often goes unrecognized in cultural and visual studies scholarship. We cannot discuss images of the laborers, for example, without acknowledging the oppressive heat and unsparing working conditions of the canyon and the desert, nor can we discuss the use of images to promote the dam's construction without acknowledging the lack of drinking and irrigation water in much of the Southwest.[10] Moreover, just as important as the physical features and immense challenges posed by the location of the dam site are the attitudes and beliefs reflected in visual and verbal portrayals of nature within Hoover Dam imagery. When viewed as a whole, these presentations reveal a homogenous narrative showcasing a rigorous scientific management approach to the land and water, one that exudes what James C. Scott describes as an ethos of "high modernism," a hubristic confidence in the linear march of American and European scientific and industrial "progress" that centers on an ever-increasing control over both nature and human nature.[11]

Although most treatments of Hoover Dam's construction have been historical, my approach in this book is primarily rhetorical; therefore, I am concerned not just with what happened, but with the *perceptions* of what was happening, the ways in which words and images worked to shape public understanding of what *should* happen, and the ways in which audiences might have interpreted particular verbal and visual messages based on their cultural and historical milieu. Consequently, I am concerned with not only what is represented but also the attitudes and values expressed in dam-related discourse. "Reclamation," for example, might have been a national imperative, but I am concerned with the ways in which producers of visual and verbal messages translated such a concept into physical reality, the means and modes of message circulation, and how and in what contexts viewers may have received those messages.

As my focus here is on imagery, I also draw from concepts in visual studies (or visual culture, or visual cultural studies),[12] an area of inquiry that accepts the visual as the focal point in the shared cultural processes by which meanings are made through the realm of representations, as well as how cultures interpret those representations.[13] The emergence of this perspective as a mode of inquiry into understanding how meaning is created through the visual originated as a way to explain the so-called visual turn of our increasingly visually oriented society. The visual turn is now well established, and although the historical privileging of the verbal is pivoting toward an embrace of the visual, I argue that this retrenchment should not be cast in terms of a struggle for primacy between image and text.[14] Instead, I show throughout the book that images and text do different yet symbiotic work in creating cultural meaning; indeed, this follows in the footsteps of some of the most important initiators of the visual turn who were, in fact, wary of any strict separation of image from text.[15]

This position is especially important for the present study because of the basic fact that visuals of Hoover Dam (and visuals generally) are routinely conjoined with verbal text or texts. Images in newspapers, books, or brochures, for example, are always accompanied by a headline, caption, or article, and often by all three. In the case of newspapers and similar documents, visuals are typically, and literally, enclosed in a verbal shell. This shell is permeable, however, as the verbal and nonverbal elements on the page freely "speak" to one another and the "message" of one may jostle with that of the other. Consequently, the reader-viewer's interpretation of the meaning of the document as a whole is one that should account for the interaction between image and text.[16] In other words, even if images cannot be said to present overt arguments, as some have claimed, arguments are apt to get attached to images in the course of their circulation (which is often a circulation of *images + texts*).[17] For example, in the case of newspapers we can see how arbitrary distinctions between the domains of *rhetoric* (verbal or written communication and persuasion), *culture* (beliefs, attitudes, and traditions), and the *visual* (photographs, illustrations, typography, page design, and so on) can create artificially delimiting categories as these three elements almost always work in concert, not in isolation. Consequently, I show throughout the book that the rhetorical force—the persuasion—of an image arises from its placement within a discursively blended visual-verbal framework and argue that attending to an image removed from its verbal, cultural, historical, and ideological context will fail to render an adequate account of how meaning is produced.

I also recognize that visuals work as a site of meaning making where social categories and differences are constructed. Examining visuals requires a critical eye toward the particular social values or ideological frames expressed by a given image.[18] As Donna Haraway and others have reminded us, it is important to acknowledge that representations produced by particular institutions privilege certain ways of seeing, which are intimately tied to ideologies such as capitalism, colonialism, or patriarchy.[19] The role of the viewer, then, is to recognize that images are directly tied to particular social power relations, thus creating the necessity to understand how images are produced or, in other words, the circumstances of their production. *Imaging Hoover Dam* thus seeks to illuminate the visual-verbal culture of the dam by emphasizing the interanimation of images and text placed in historical, cultural, and ideological contexts, which presupposes that visuals work in combination with some type of text, context, and audience, each encouraging different interpretations. This perspective accepts that images are always embedded in a range of discourses (visual, verbal, and cultural) that continuously intersect and influence one another, and that they are typically integrated for some persuasive purpose.

In many ways, then, this book is not just about the dam, but about a coordinated visual-verbal messaging schema controlled by business, government, and news media that shaped the public's perception of the dam. It argues that the cultural mythology of Hoover Dam was in fact invigorated by an array of visual representations depicting the dominating and awe-inspiring facticity of the undertaking while at the same time creating a sacred narrative of the dam and those individuals and agencies that built it, one that, through its scale and heroic narrative of overcoming the greatest of natural obstacles, has exemplified the "powerful and original America" that Jean Baudrillard described as a "Utopia made reality of a society which . . . is built on the idea that it is the realization of everything the others have dreamt of—justice, plenty, rule of law, wealth, freedom: it knows this, it believes in it, and in the end, the others have come to believe in it too."[20]

Ultimately, it is the aim of this book to use the exploration of Hoover Dam imagery as another step toward understanding the role of all images circulating within our culture, from the quotidian to the great masterworks, as part of a broad discursive field that both constitutes and reflects the values, beliefs, and aspirations shared by a community or even a nation.

1

Nature, Culture, and the Divine Right of Transformation

The construction of Hoover Dam was an achievement that owes its realization to a long and rich history of Western religious, political, and cultural imperatives; thus, it is vitally important to establish this framework at the outset if we are to understand how and why certain images were created and circulated at particular times in the dam's history. To that end, this chapter provides some contextual background regarding the philosophical and ecclesiastical antecedents of American attitudes toward nature, while in the next chapter I take up the political maneuvering that paved the way for the Hoover Dam project itself. These concerns undergird the many ways in which ideologies of land reclamation and "wasted" natural resources were visually and verbally portrayed to the public. This also helps to establish how traditional American attitudes regarding nature were a driving force in both the dam's construction and its resulting visual and verbal representations. In this and the following chapter, I frame the broad sweep of exigencies that led to the building of the dam, which ranged from a nature/culture dichotomy infused by traditional religious doctrine and a headlong rush toward modernity, to political and legal wrangling, to the physical environs of the Southwest and the migration patterns of Americans in the early twentieth century. The remaining chapters, then, center on the various ways in which visual images were deployed (or not) in response to many of those exigencies.

I should, however, make a distinction here at the outset between what I argue are two different notions of American culture. The first is what, for our purposes here, can be called the "weak" version, one that stems from cultural and visual cultural studies, of which its contemporary notions are indebted to the foundational work of Raymond Williams and Stuart Hall. I describe this as weak not because it is lacking in any sense, but because it is the traditional notion of culture that is familiar to many, yet is rather muted in its importance for my discussion of Hoover Dam. The second, what I will consider the "strong" version, understands culture as being in a reciprocal relationship with the natural world, one that is a product of the confluence of religious doctrines, agrarian philosophies, and the long-standing drive to use science and technology to control and order nature.

The term *culture* itself has often been described as one of the most complex words in the English language, one that generally encapsulates the shared traditions and processes that allow an individual to make sense of themselves within a community, especially when compared to other communities. There are untold variations of the term, but in looking at the most cited definitions, a few ideas emerge that are common to all. First is that culture is a strictly human endeavor. Second is that it involves behavioral patterns that are shared and learned within a particular community of people. Third is that these behaviors are transmitted and passed along to other groups, including subsequent generations. Fourth is that culture involves human symbols and artifacts. If nothing else, culture seems to comprise these elements.

The term and how those interested in visual and cultural studies understand it today have two principal origins. The first is from Mathew Arnold via Clement Greenberg. Arnold's 1869 series of essays *Culture and Anarchy* discusses those two titular terms as binaries and argues from the bourgeois perspective that culture is akin to high culture, all the "best" art and literature of a society produced by its greatest and most accomplished members. Greenberg furthered Arnold's perspective in his influential 1961 essay "Avant-Garde and Kitsch" by contrasting Arnold's notion of (high) culture with the lowbrow tawdriness of mass-produced kitsch consumed by the underclasses and uneducated. The second begins with anthropologist E. B. Tylor, whose 1871 book *Primitive Culture* moved away from the importance of place, time, and high versus low cultural taste and instead sought to describe how humans create "non-natural" artifacts such as beliefs, art, morals, laws, customs, and so forth that hold meaning for particular groups.[1] Many people have since advanced a variety of iterations of these concepts, but recent trends in cultural studies and visual culture adhere to Tylor's conclusions more than Arnold's.[2]

The way that I use *culture* in this book, however, centers on an American tradition of land transformation, one that embodies an ethos of modernity eager to use science and technology to transform or order nature into utilitarian functionality. This perspective frames the Hoover Dam's construction as a beacon of modernism and the apogee of man's long struggle against nature, one that spurred a frenzy of dam building in the United States and around the world. It also views Hoover Dam as another in a long line of liturgically motivated endeavors to use technology to re-create or reclaim the Garden of Eden for the profit of humankind, a process I term the "divine right of transformation." Unlike most definitions, however, it also acknowledges the place of nature in the development of culture and references a principally American way of life that is largely the consequence of the unique intricacies of the history, geography, and resources of North America. This view is one that is also permeated by Protestantism and agrarianism, two notions that have traditionally shaped American attitudes and values regarding the role of human labor in relation to the land.

This conception of an American culture driven by a divine right of transformation is also influenced by ideas such as Frederick Jackson Turner's Frontier Thesis as articulated in his highly influential 1893 essay "The Significance of the Frontier in American History." Although Turner's Frontier Thesis has been the subject of much criticism—and occasional outright dismissal—for his narrow view of the roots of American cultural development, and particularly for his failure to consider the exploitation of natural resources, or the role that women, Native Americans, African slaves, and other groups played in the growth of the nation (among other shortcomings), his broader conception is still important. While some would contend that the *idea* of the wilderness and frontier was a more powerful shaper of American attitudes than the actual frontier, Turner's argument that the American West molded the American character—a character that embodies the spirit of rugged individualism and an ability and determination to tame the land—spurred an intellectual shift in our consideration of the role of the West in the nation's history.[3]

Turner writes in the opening statement of his essay, "Up to our own day American history has been in a large degree the history of the colonization of the Great West. The existence of an area of free land, its continuous recession, and the advance of American settlement westward, explain American development." To him, as generations of people moved west and settled the land, they endured new hardships and conquered the great American wilderness, what he describes as "the meeting point between savagery and civilization." As a result of their experiences in the wilds of the frontier, these settlers forged from their struggles a uniquely American set of characteristics highlighted by innovation, plainspokenness, and perseverance that served to strengthen democracy by weaning Americans from the genteel influences and institutions of Europe.

Turner contends that the American people and its institutions are a unique by-product of the adaptations made "in crossing a continent, in winning a wilderness, and in developing at each area of this progress out of the primitive economic and political conditions of the frontier into the complexity of city life." As this process unfolded, forming what Turner calls a "composite nationality" of easterners and settlers, the inimitable challenges of settling the American West served as a crucible in which Americanism was wrought by the frontier, a circumstance that existed uniquely at this particular intersection of civilization and wilderness. To him, this was *the* crucial circumstance that set Americans apart from Europeans.[4] As Turner's thesis suggests, any conception of an "American" culture must take into account the dichotomous relationship its people have had with nature, one that has been in conflict with, exploitative of, yet strongly shaped by and dependent on the physical environs of North America.

The Cultural Construction of Nature

The idea that nature influences culture and that particular environmental experiences have worked to shape cultures is richly articulated in the work of environmental historians who suggest that it is impossible to understand our past, our history, without taking into account the world of plants, animals, water, soil, and weather (among other things). Environmental history's thesis is simply that nature has shaped human history as much as human history has shaped nature. This perspective holds that ecological events influence historical events, and that the distribution of natural resources shapes politics, public policy, and wars; influences migrations of people; and dictates the rise and fall of empires.[5] Environmental history also seeks to examine the ecological past as an influence on the political and economic present while binding together concerns of biology, the environment, and technology.[6] This perspective considers the modes of production and roles of domination that rule twentieth-century democracies as the drive to remake nature for the purposes of utility and participation in the material economy impels people to use technology in ever more invasive ways. This urge to use what Donald Worster describes as "instrumental reasoning" is driven by capitalistic ideologies and is the way in which we look at nature with utility as our only concern. The US government's approach to land reclamation in the American West, for example, has often been summarized as a form of instrumental reasoning: the scientific, utilitarian conversion of natural resources into wealth and power.[7]

A persistent theme for environmental historians is also the notion that nature is not necessarily a pristine entity set aside from human interaction, and that the idea of an autonomous nature "out there" free from the influence of humanity makes little sense. In fact, as long as bipedal beings have walked the earth, they have left their imprint on the environment. For hundreds of thousands of years, human and humanlike species have left their ecological imprint on nature by hunting animals, burning grasslands, constructing cities, building dams and irrigation canals, creating tools, and engaging in agricultural and animal husbandry practices. When viewed from this perspective, the boundaries between nature and culture appear less clear than they might at first seem.

Of course, what we name things is very important and often controversial, and terms related to the environment, such as *nature* or *wilderness*, for example, have been particularly problematic. We can no doubt recognize places that exist that are wilderness and nature; however, along with those terms comes at least several centuries worth of cultural baggage. What, for example, constitutes a wilderness? The received notion of wilderness implies a distinct lack of humans and human activities. This notion is even codified in the Wilderness Act of 1964 as an area "in contrast to those areas where man and his own works dominate the landscape, is hereby recognized as an area where the earth and its community of life are untrammeled by man, where man himself is a visitor who does not remain."

But by burning fossil fuels, for example, humans produce carbon dioxide that reaches every corner of the globe and, many would argue, "trammels" the biota and landscapes found there. Do carbon dioxide and innumerable other gases resulting from human industry not "dominate the landscape" in every conceivable aspect, even where humans do not live? Furthermore, some environmental historians argue that wilderness (the very idea so important to Turner's theory of Americanness) is not the last sanctuary left untouched by humans and uncontaminated by civilization, but simply a conjured by-product of civilization's desire to feel that such places do in fact exist.[8]

Like wilderness, most people would agree that there is something "out there" that we generally refer to as *nature*. But being rooted in language, concepts of nature and "natural" have changed radically throughout history and have had vastly different meanings, connotations, and values across cultures. As with *wilderness*, the linguistic signifier *nature* has a long history of volatility as there have been myriad contestations of the term during the medieval period, the Renaissance, the empiricism of the seventeenth century, and right up to today, all in an attempt to describe a culturally determined variable of what is widely considered to be an empirical and denotable entity.[9]

Recent trends in environmental history contend not only that nature is not pristine, but that it is not even a constant, and that there could in fact be different kinds of nature. Moreover, even the distinction between urban and rural nature (or the lack of urban nature) is a misconception that some argue has pervaded nature writing and environmental history.[10] Americans have long understood *the city* and *the country* as isolated, separated places without much connection. Consequently, everyone from urban dwellers to farmers to national parks directors have carefully and methodically planned and separated—in both mental abstraction and actual practice—urban, rural, and wilderness landscapes. However, most notions of city and country, or urban and rural, share an underlying and deeply problematic assumption, namely, that those places have some essential or innate difference.[11] Thus, we somehow believe that it is possible to make distinctions between urban and rural landscapes when, in fact, every distinction between the two is either capricious or arbitrary.[12]

The host of permutations of nature versus culture, or nature-culture (such as Bruno Latour would describe), or "it's culture all the way down," or any combination of the sort often causes fits of hand-wringing for those attempting to discuss either the application or theorization of these terms. In *The Idea of Culture*, British literary critic and philosopher Terry Eagleton wryly brings these two terms, *nature* and *culture*, together and explores the meaning of the word *culture* alongside what is often considered to be its equally complex opposite—*nature*. We can draw on Eagleton's work to better understand the assumptions and reciprocity with which these terms might intermingle as a stepping-stone toward our entry into the cultural and historical antecedents of Americans' attitudes toward nature that eventually lead up to the construction of Hoover Dam.

Eagleton suggests that there exists a dialectical relationship between the artificial and the natural world: the things we do to the world and the things the world does to us. To him there is no need to deconstruct any opposition between nature and culture because they are already deconstructions—epistemologically realist notions with a constructivist dimension. In other words, there is a real world out there constituted of raw materials that must, however, be worked upon by humans to take any significant or meaningful shape for us.[13] That all seems straightforward enough, but things are not quite as simple as they might appear. This is the point at which most people end, but it is exactly the point where Eagleton starts.

There is a continual reciprocity between nature and culture that has been pointed out by Judith Butler and others; however, Eagleton adds treatments of both politics and religion that are particularly relevant for our purposes here. There exists a curious yet important self-reflexivity in that humans are products of nature, but unlike other products of nature, we can shape and mold ourselves, we are "self cultivators"; thus, our civilized condition is actually quite "unnatural." The terms *civilization* and *culture* are not interchangeable but are actually opposite terms borne of the Enlightenment, where culture is "tribal" rather than cosmopolitan. Culture lives at the primitive level of social order, where it is a universalizing force, but it is also radical and destructive at the same time with a built-in, ironically self-destructing mechanism in that the things that allow cultures to flourish are the very things that work to tear them down. The so-called counter-cultures are actively working to frustrate the corporate cultures, and any idyllic notion of cultural unity is always at odds with an inherent conflict of cultures.[14] Further, there are destructive forces within cultures themselves—desire, domination, violence, vindictiveness—that spring up from a society's interface with nature and other cultures.[15]

Like Stuart Hall, Eagleton calls for us to bring into sharper relief a forceful determiner of both nature and culture: politics. Eagleton suggests that political interests govern cultural ones and thus define our notions of culture, nature, and humanity. He reminds us that for his mentor Raymond

Williams, "what matters most is not cultural politics, but the politics of culture. Politics are the condition of which culture is the product."[16] For Eagleton, though, politics is still not enough. As well as being heavily influenced by Marx, Williams, and the psychoanalytic work of Slavoj Žižek, *The Idea of Culture* is quite reflective of Eagleton's forays into theology. Consequently, in a comparison of T. S. Elliot and Williams, he speculates that it is, in fact, religion that is the most powerful ideological force that humanity has ever conceived.[17] To him, religion as ideology is powerful enough to bond people where political and cultural unity has failed. He writes, "No form of culture has proved more potent in linking transcendent values with popular practices, the spirituality of the elite with the devotion of the masses. Religion is not effective because it is otherworldly, but because it incarnates this otherworldliness in a practical form of life. It can thus provide a link between absolute values and everyday life."[18] It is within this nexus of politics and religion—the religiously invoked fervor to "reclaim" lands and waterways (absolute value) linked with the need for electricity, safety, and irrigation and drinking water (everyday life)—that we begin the story of America's drive to build the biggest dam in the world.

The Divine Right of Transformation

As the name and promise of Imperial Valley spread across America, it began stoking the passions of anyone who saw profit and opportunity in the vast parched southwestern lands formerly known as the Colorado Desert. The Imperial Valley's early development was based on high-risk real estate gambits that were in turn based on the uncertain prospects of bringing to the area irrigable and potable water. This rampant speculation, however, ignited the imaginations of engineers, real estate investors, railroad barons, politicians, and journalists who saw the potential for vast tracts of farmland and burgeoning cities resulting from Western reclamation projects.[19] The *Las Vegas Evening Review-Journal* wrote in 1932, "Nothing in this decade has captured the public imagination as has this daring attempt of man to tame a mighty river that has long defied him."[20]

Regardless of its potential, luring residents to the harsh, flood-prone, and as-yet-undeveloped Colorado Desert, and persuading a reluctant government deep in the throes of the Great Depression to spend tens of millions of dollars on a giant dam—what some saw as a boondoggle of monumental proportions—would require a massive public relations effort.[21] Although often focused on an uncertain economic climate and the promise of a new start in life, the underlying success of the appeals for the necessity of the project hinged on visual and verbal messages that tapped deep-seated cultural and religious values that have historically driven Americans' use of technology to transform the land—a devoutness rooted in using technology to remake or "reclaim" nature in man's interest.

To "reclaim" suggests that the lands are wild, untamed, and need to be controlled. But more importantly, it also suggests that those lands were once "claimed." The notion of reclaiming lands is one that has a long history in Western thought, extending back through Jefferson and Milton, and to at least early Protestantism. This has been described as a grand narrative of declensionism in which humans have lived in a state of moral deterioration and a continuous falling away from the earthly paradise of the Garden of Eden, and therefore all of Western history is an attempt to re-create or recover Eden and transform the earth into a garden paradise.[22] Carolyn Merchant, one of the scholars most associated with this thesis, argues that a "Recovery narrative," in which humans seek to restore or re-create the earthly paradise, has shaped human history, particularly since the scientific revolution of the seventeenth century. For her, this Recovery narrative is *the* compelling mainstream narrative of modern Western culture. She writes, "Aided by the Christian doctrine of redemption and the inventions of science, technology and capitalism, the long-term goal of the Recovery project has been to turn the entire earth into a garden."[23]

The metaphor of the garden has been a fixture in the attempts to describe humans' relationship to nature.[24] Leo Marx, in *Machine in the Garden,* describes the romantic pastoralism of the American West as an unexplored garden that was eventually overwhelmed by technology in man's attempt to cultivate it for consumption. Daniel Philippon has written that many prominent American nature writers have used metaphors of garden (Mabel Wright), as well as park (John Muir), and utopia (Edward Abbey) that have conveyed environmental values through narratives about the human relationship to nature.[25]

Biblical references such as the Garden of Eden and the story of Noah's Ark were also regularly attached to the dam project, as well as to the beginnings of numerous farming towns and other municipalities throughout Southern California.[26] The prevailing notion at the time was that if engineers could just find a way to bring a steady flow of irrigation water to the Imperial

Valley without the threat of floods destroying everything, then they had the potential to create an "Eden" of the Southwest. Recalling Isaiah 35:1–2 ("The desert and the parched land will be glad; the wilderness will rejoice and blossom. Like the crocus, it will burst into bloom; it will rejoice greatly and shout for joy"), settlers from the Spanish to the Mormons to colonial Americans sought to create in the American West a vast garden of biblical proportions. In fact, the Bureau of Reclamation itself was established in 1902 by President Theodore Roosevelt to, according to him, "make the desert bloom."

US government publications, too, drew on biblical references to dramatize reclamation projects. One early 1930s publication sponsored by the US Bureau of Reclamation, for example, is rife with biblical language, such as the heading "Man Tempts the Colorado," with the subsequent text including a variety of references to an "Eden" in the Imperial Valley. The publication describes the development of agricultural lands near the Colorado River as the "story of their creation" and traces its transformation from unusable and "wasted" lands to productive lands created by an "imaginative white man."[27] It also goes on to explain the "Genesis of Imperial Valley," in which the river, in "historic time," flowed of its own accord, regularly spilling over its banks and returning to its original channel, a cycle that was of "little concern to man until steps were taken to develop the Imperial Valley for agricultural purposes—then the inundations became a menace to life and property."[28]

Newspapers also wrote of the bounty that awaited those who brought water to the West, and often with similarly ominous descriptions of the ever-threatening Colorado River. The *New York Tribune* in 1922, using illustrative photographs of the proposed dam site and a photograph of a grapefruit orchard in Arizona, wrote that the Imperial Valley, "once an almost impenetrable desert . . . is now the land of milk and honey—but with a constant menace of destruction hanging over it like the proverbial sword of Damocles."[29] Robert L. Duffus wrote in the *New York Times* in 1924 that "once the danger of floods is removed, the Southwest will enter upon a period of growth and prosperity which will be as important in its way as the gold rush into California. The 'Great American Desert' of the old geographies will at last literally blossom."[30]

The divine right of transformation, or the scripturally endorsed use of technology-assisted labor to mold the landscape to the needs of humankind, has a long history in Western culture, going back at least to the medieval and Renaissance periods, in which the theological notion of using technology as a means to re-create the worldly perfection of the Garden was sanctioned by God and even obligatory.[31] The Benedictine order of monks felt that manual landscaping and liturgical practices were concomitant. Hugo of St. Victor taught that the mechanical arts, such as navigation, hunting, and agriculture, and the liberal arts, such as grammar, rhetoric, and geometry, should be learned and practiced in harmony and on equal terms with each other.[32] John Milton, too, writes in Book 10 of *Paradise Lost* (1667) that although humankind's freedom has already been restricted by the Fall, nature will surrender to man under the pressure of reason and technology; the earth and heavens will be within man's domain if individuals think and act separately and for God.

In America, this transformational prerogative has strong religious roots that include both Calvinist and Puritan branches. John Calvin taught that all men, rich and poor, must work, and to work was to fulfill the will of God. In a major departure from traditional Christian teachings, Calvin maintained that unlimited profits were to be gained through reshaping the world in the image of the kingdom of God. To him, God provided technological and mechanical means unto man, and it was man's obligation to use them upon the earth to please God. The Puritan colonists fleeing religious persecution brought with them a faith-based work ethic and the desire to create for themselves a new Edenic state. According to their creed, if they failed, or if hard times were upon them, they were simply not up to the task that God had laid before them, and it was incumbent upon them to then create better technology and work even harder to transform the land for more efficient and productive utility.[33] According to Protestant teachings, it is through hard work and success that men reap both worldly and heavenly rewards and demonstrate that they are among the Select who are predestined to receive salvation.[34] The profits from all of this work, however, are intended not for personal enrichment but to be reinvested to create even more opportunities for others in less fortunate positions. For Protestants, work (assisted and enhanced by technology and machinery) was the core of a moral life.[35] These doctrines created a culture that encouraged the use of technology and mechanization to exploit the land to maximum benefit until the end of time.

To many American colonists, then, the natural world was unfinished and simply awaited useful improvement. Furthermore, the process of technological advancement, as well as the domination of nature and native peoples, was nothing more than an inevitable process. This type of thought was most clearly exhibited in the expanding philosophical and political notions of the agrarian society, the prevailing doctrines of which were that agriculture was the only source of true wealth, man had a natural right to land, ownership of land makes farmers independent and provides the means for social status and dignity, and government should be dedicated to upholding these ideals and the interests of the farmers.[36] Although the beginnings of agrarianism were present in Benjamin Franklin's writings of the 1750s, the adherence to this kind of idealism is most widely recognized in the writings of J. Hector St. John Crèvecoeur and Thomas Jefferson.

Jefferson was particularly instrumental in transforming the notion of the simple farmer from the old image of laboring peasant to the new symbol of republicanism. His poetic idealization was of the farmer transforming his own little plot of fallow land into a beautiful and life-affirming garden—something in between the wild and uncultivated wilderness of the Americas and the human-built metropolises and industrial complexes of England. To Jefferson, farming was the foundation for a moral and virtuous life. In a 1785 letter to John Jay, Jefferson wrote, "Cultivators of the earth are the most valuable citizens. They are the most vigorous, the most independent, the most virtuous, and they are tied to their country and wedded to its liberty and interests by the most lasting bonds."[37] Later, in a letter to George Washington in 1787, Jefferson wrote, "Agriculture . . . is our wisest pursuit, because it will in the end contribute most to real wealth, good morals and happiness."[38]

This notion of an agriculturally based society sustaining the citizenry's moral rectitude was not only a philosophical stance but also a political one. In his 1776 draft of the Constitution of Virginia, Jefferson argued that every landless adult white male should be given fifty acres from the public domain,[39] and in a letter written to James Madison in 1785, Jefferson famously wrote that "small landholders are the most precious part of the state."[40] Elsewhere, in a section of his *Notes on the State of Virginia* (1787) titled "Manufacturers," Jefferson contended that America should build its economy on agriculture—a nation of independent farmers—rather than on manufacturing. He wrote, "Those who labor the earth are the chosen people of God, if ever he had a chosen people."[41] For Jefferson, this notion of agrarian idealism was the material embodiment of what had been to Europeans merely a utopian fantasy. Moreover, the vast expanse of undeveloped land that awaited beyond the Allegheny Mountains held the potential for an agrarian ideal so boundless that it far exceeded any previous conception of agrarian societal transformations.[42]

Jefferson, like Thoreau and others, mostly in reaction to the seemingly inevitable encroachment of technology and mechanization into society, eventually acquiesced to the presence of such things as locomotives and hydro-powered textile mills, but only so long as these technologies treaded lightly on the landscape and the majority of the land remained in the hands of farmers, an arrangement that Leo Marx famously describes as the "machine in the garden."[43]

The Enlightenment era also brought about technological and cultural changes that were conducive to the development of a sense of what Marx refers to as "the pastoral," or "the widespread tendency to invoke Nature as a universal norm; the continuing dialogue of the political philosophers about the conditions of man in a 'state of nature'; and the simultaneous upsurge of radical primitivism . . . on the one hand, and the doctrine of perfectibility and progress [of nature] on the other."[44] By the time Crèvecoeur printed his *Letters from an American Farmer* in 1783 and Jefferson his *Notes on Virginia* in 1787, the pastoral image had been transformed into an all-encompassing ideology in America.[45]

Despite the fact that the pastoral was an important, albeit highly malleable, concept in early American life, it was one that was burdened with an intrinsically conflicting dualism: on the one hand was the desire to use technology to transform the land, and on the other was the radical primitivism and longing for the mythic simplicity of our forefather's humble relationship with nature.[46] An article titled "Pastoral of the Far West" that appeared in the *Los Angeles Examiner* on July 16, 1929, encapsulates the hopes of many that these tensions could be alleviated and, in fact, united by embracing water reclamation. Showing a photograph of a herd of dairy cows standing in the shade of a large tree, the article seeks to provide a vision of a future of the Southwest "transformed by irrigation water from the Colorado River made possible by Hoover Dam and the All American Canal." The caption states that the photograph is "not a painting, but a glimpse of a contented herd on an Imperial farm—which once was only a desert."

FDR and the Discourse of Man Dominating Nature

On September 30, 1935, President Franklin D. Roosevelt dedicated Hoover (Boulder) Dam in a live radio address to the nation from atop the newly completed structure. A central premise of Roosevelt's dedication address was the notion that awe and veneration should be reserved not for nature, but for man's ability to control nature through technology. In his address Roosevelt emphasized the economic benefits of land planning by means of a contrast with the wild unruliness of nature left to its own devices. To him, labor and materials make wealth, whereas unplanned, uncontrolled nature constitutes wasted resources: while the pre-dam Colorado River "running unused into the sea" was a squandered resource, once the river began working to generate electricity (and wealth), the Colorado became "a great national possession." This is clear at the outset of his speech when he states, "Ten years ago the place where we are gathered was an unpeopled, forbidding desert. In the bottom of a gloomy canyon, whose precipitous walls rose to a height of more than a thousand feet, flowed a turbulent, dangerous river. . . . The site of Boulder City was a cactus-covered waste. The transformation wrought here in these years is a twentieth-century marvel." Later he adds, "We know that, as an unregulated river, the Colorado added little of value to the region this dam serves. . . . For a generation the people of the Imperial Valley had lived in the shadow of disaster from this river which provided their livelihood and which is the foundation of their hopes for themselves and their children."

Roosevelt's words resonated with those in attendance, as the *Los Angeles Times* reported the next day that his remarks were "accepted with murmurs of 'Amen'" by the crowd. These sentiments, however, were not new for Roosevelt. At an August 1934 speech in Green Bay, Wisconsin, Roosevelt invoked the spirit of the pioneer who struggled against nature to make it a "more useful servant," contending that our spirit should propel us to "master our environment. . . . The old frontier is occupied, but . . . science and invention and economic revolution have opened up a new frontier."[47] All of these descriptions frame the landscape as useless and valueless because it provides no tangible economic benefits to the people already living in the region and no benefit to those quickly populating the Southwest.

Regardless of its inherent duality, the impulse toward agrarian and pastoral thought in America was not lost on Roosevelt when he gave his dedication address at Hoover Dam. Paralleling Jeffersonian agrarian philosophies,

Roosevelt had previously said in response to concerns about the balance between agricultural and business interests that it was the farmers who should achieve economic parity with industry and that irrigation and soil conservation practices were the way to achieve that goal.[48] It seemed to him that greatly increased irrigation would open vast areas of unclaimed land to industrious farmers who could fulfill Jefferson's dream of self-reliance and virtue by tilling their own soil. Because of this, Boulder Dam was a project through which Roosevelt could champion a variety of ideologies and practicalities: farming as a national imperative, public-private partnerships, water reclamation, jobs for the unemployed, and the use of technology to render "wasted" nature useful. Roosevelt wanted farmers to learn to "use every square mile of the United States for the purpose to which it is best adapted,"[49] an approach that he saw as an elixir for the national state of economic depression.[50] In fact, Roosevelt championed hydropower projects throughout his presidential election campaign and referred to "four great government regional units" in his dedication speech: Hoover Dam on the Colorado River, the Muscle Shoals project on the Tennessee River, the St. Lawrence River dams and opening the Great Lakes to deep-water navigation, and Columbia River dams designed to provide hydropower and to allow large-scale shipping up the Columbia all the way to the mouth of the Snake River. In 1932, R. L. Duffus summarized Roosevelt's "water-power plank" in the *New York Times* this way: "When Governor Roosevelt in his present speech-making tour has wanted an illustration for his water-power plank he has found it in four great projects, all of them involving vast expenditures of public money, all of them advocated because they promised to return far in public benefit that they cost in money, and the sum total of them outweighing almost any other single constructive enterprise of mankind except the building of the roads and railroads."[51]

THE TONE OF ROOSEVELT'S ORATORY was not unusual for early twentieth-century America. The pervasive discourse of man dominating nature in the early 1900s and the resulting "golden age" of dam building throughout the American West from the 1930s to the early 1950s has been well documented.[52] This discussion highlights a long-held, religiously sanctioned Western tradition of striving to improve the land through human endeavor augmented by technology. As Leo Marx has observed, "to understand the American consciousness in this period, the key image, as Tocqueville noted [in *Democracy in America* (1840)], is the 'march' of the nation across the wilds, 'draining swamps, turning the courses of rivers, peopling solitudes, and subduing nature.'"[53] Likewise, in John Stuart Mill's commentary on Tocqueville's writings, he argues that, to Americans, technology and machines are powerful symbols of progress and the enlightenment of mankind. Machines—steam locomotives roaring across the landscape, for example—inspire admiration among men regardless of class or education.

It is these ideas that have shaped American notions of nature and its role in American culture. Hoover Dam underscores these notions through its depiction as a titanic clash of human labor and technology against the "menacing" Colorado River, the latest and greatest in a long line of man's struggles against a sublime nature, and a testament to America's divine right of transformation and the inevitable march of American greatness. The narrative of Hoover Dam—man using technology to tame nature for the betterment of the country—fulfilled long-held American religious and cultural beliefs. Yoked together and displayed in the form of utopian visual and verbal assemblages, these narratives formed a cultural imaginary of Hoover Dam, a shaping mechanism for regional and national identities. Hoover Dam was another step toward American manifest destiny and an undertaking that was projected to transform the land for the betterment of humankind by fulfilling the tenets of Calvinism, Protestantism, capitalism, and high modernism. Without the dam, many said, the Southwest was fated to be the "cactus-covered waste" that it had always been.

2

Natural Disasters and Political Adversity

Whether because of extended drought or raging floods, in the years leading up to and through the start of construction, the erratic nature of the Colorado River—what Philip L. Fradkin calls "the most used, most dramatic, and the most litigated and politicized river in this country, if not the world"—jeopardized the growth of the Southwest.[1] Although the combination of the Colorado's constant threat of flooding and its potential to provide irrigation and drinking water to the parched Southwest was certainly the main impetus behind building the dam, numerous other significant national events underscored both the necessity and controversy of the project and framed the context in which the dam's construction took place.

Part of the reason that Hoover Dam captured the nation's imagination so quickly and has been fixed in Americans' collective memory for so long is because it was built and documented during a time of deep national anxiety: the Great Depression of the 1930s, the most severe social and economic shock America had experienced since the Civil War. The Stock Market Crash of 1929 resulted in a 90 percent decrease in the market's value. The crash was followed by an epidemic of bank failures caused by risky investments and a lack of cash reserves. The public, in fear of losing their savings, began withdrawing cash assets, which, in turn, caused many banks to call in loans and foreclose on mortgages to cover the outflow of cash. To add to the depth of the economic problems, poor farming practices, coupled with a nearly decade-long drought that afflicted much of America, resulted in unprecedented agricultural failures. The chain reaction set in motion by the stock market crash (but preceded by years of misguided business decisions and governmental policies) resulted in a severely depressed economic climate throughout the 1930s. Millions of American families became homeless and dependent on bread lines and soup kitchens as food shortages and near starvation were common themes for Depression-era Americans.[2]

The market crash and resulting depression also ushered in an era of political dissatisfaction leading to Franklin D. Roosevelt's resounding defeat of Herbert Hoover in the election of 1932. In the following years, one of the most ambitious ventures of the New Deal, the newly formed Tennessee Valley Authority (TVA), began construction on huge public works projects throughout the midwestern United States. But despite government efforts, the Great Depression was continuing to buffet the US economy, as unemployment fluctuated between 15 and 25 percent, while the dust bowl drought conditions persisted across the prairie states, causing tremendous ecological damage and agricultural losses. At the same time, the Farm Security Administration (FSA) was in the midst of lending more than $1 billion to US farmers as part of Roosevelt's second New Deal.

But for the Hoover Dam project, two things were even more important than the national milieu of the Great Depression: the mass migration of people to the Southwest and particularly to Southern California, and the agricultural potential of Imperial Valley. As a westward-moving US population began to settle in the salubrious regions of the California coast, Los Angeles developed from a settlement of less than ten thousand people to a booming metropolis of over one million and growing. While Los Angeles city officials sought to secure supplies of drinking water and cheap electricity, residents of Imperial Valley sought protection from the yearly flooding of the Colorado that wrecked irrigation canals, swamped farmland, destroyed businesses, and at one point threatened to inundate the entire southwestern United States.[3]

To those living in the Imperial Valley, the "untamable" Colorado was an ominous and omnipresent danger. In 1925, Congressman Phil D. Swing, a resident of Imperial Valley, explained in an article in the *Santa Ana Daily Register* that the principal threat to the valley and its people was the "annual flood menace." He described the sixty-five thousand people living in the Imperial Valley as "pioneers, drawn from all the States in the Union . . . with willing hands and determined minds they set out to make something worth while out of what appeared to be a desert waste. They fought a good fight and conquered the desert, they builded [sic] their cities well, they created a magnificent community, and yet today these people are face to face with the menacing threat of destruction by flood."

Although the Colorado's eons-long history of flood-then-drought cycles had frustrated agricultural efforts of white and European settlers, it was something with which indigenous peoples had learned to largely coexist. The Mormon settlers of Utah are frequently given credit for developing the first large-scale irrigation projects in the American West, but that is not the case. It has been well documented that for many years—nearly a thousand years before the Spanish and white settlers arrived—indigenous peoples such as the Hopi and the Tohono O'odham of present-day southeastern Arizona and northwestern Mexico were engaged in water irrigation projects of their own, some even on a very large scale covering many square miles. The Hohokam of present-day Phoenix, Arizona, and the Anasazi of northwestern New Mexico also built extensive irrigation canals and used temporary dams made of sticks, rocks, and branches to impound the waters for intensive, seasonal agriculture. Many of the abandoned canals of the Hohokam, some of which were thirty feet wide and ten feet deep, were reused later by white settlers, and a few are even in use today. The main difference, however, between those projects and the US government's water reclamation plan was that the indigenous peoples of the American West did not endeavor to dominate the landscape with their technology to attain hegemonic power and wealth, but rather sought to achieve a meager coexistence with the land.[4]

The Alamo Canal Scheme

Throughout the late 1800s and early 1900s, various privately financed schemes were implemented to divert Colorado River water to the Imperial Valley, the huge area of open land just north of the California-Mexico border. This expansive region (which was, in fact, the Colorado Desert) lies below sea level, making it ideal for rerouting water down into the valley, a practice that has been employed worldwide dating back thousands of years to indigenous Sumerians and Akkadians in the Tigris-Euphrates Valley of ancient Mesopotamia, as well as to tribes of western North America. Unfortunately, this unique topographical circumstance also made the Imperial Valley quite susceptible to flooding.

In 1900, investor and real estate speculator George Chaffey put into motion a series of events that would culminate in the worst flood in the history of the Southwest, an exigency that would prompt serious federal and state government efforts to build Hoover Dam.[5] Chaffey, president of the California Development Company, paid $150,000 to have engineer Charles Rockwood build a canal for farming and irrigation that would carry Colorado River water to settlers in western Arizona and Southern California.[6] Furthermore, in order to promote more settlement to the area, Chaffey replaced the gloomy-sounding Colorado Desert and Salton Sink with the more palatable and marketable name "Imperial Valley." As the irrigation canals grew, so did the population, as families, businessmen, and farmers migrated westward on the promise of economic opportunity.[7]

In 1902 Rockwood and Chaffey formed the Imperial Land Company to develop towns and sell land parcels in Imperial Valley, but after a botched attempt to build a canal south into Mexico, Rockwood bought back controlling interest in the California Development Company and ousted Chaffey.[8] Between 1901 and 1904, Rockwood's company had irrigated roughly 250,000 acres, while nearly ten thousand settlers came to the area to farm. But as the Colorado River flooded repeatedly between 1905 and 1909, threatening the newly named and freshly populated Imperial Valley with destruction, one project in particular, the Alamo Canal, would prove to be especially disastrous.

Connecting the existing Colorado River with the ancient, silt-filled Alamo riverbed, and fitted with wooden headgates designed to control the flow of water, the Alamo Canal provided an effective and cheap way to divert Colorado River water into the Imperial Valley.[9] But because of constant silt buildup and the ever-declining flow of river water for irrigation, the California Development Company decided to build a larger headgate a few miles south of the US-Mexico border that would more directly distribute water to the valley. However, the spring of 1905 brought a devastating flood that caused the Colorado River to overflow its banks, destroying many of the shoddily built wooden headgates and levees that dotted the riverbanks, including the Alamo headgate. When the lower Alamo headgate gave way, ninety thousand cubic feet of water per second poured through a 160-foot-wide channel into Alamo Canal and on into Imperial Valley. For months afterward, the river cut through the desert at the rate of more than a mile per day. Geologists predicted that, if left unchecked, the Colorado River would be irrevocably rerouted, destroying the city of Yuma, Arizona, and permanently flooding parts of Southern California and Arizona in the process.[10]

Rockwood, with the imminent destruction of thousands of square miles of the American Southwest unfolding before his eyes, appealed for help to wealthy New York businessman E. H. Harriman, president of the Southern Pacific Railroad. In exchange for control of Rockwood's business enterprise, Harriman agreed to attempt a massive engineering effort to put the river back within its original banks. But by April 1906 the late spring floodwaters of the Colorado were approaching their peak and six billion cubic feet of water per day was surging through the now half-mile-wide Alamo headgate crevasse, submerging under some sixty feet of water the houses, farms, and businesses in the heart of Imperial Valley. Unfortunately for Harriman, this newly forming "Salton Sea" was now rising at a rate of over seven inches per day and threatening to forever alter the American Southwest.

By August of 1906, after eighteen thousand tons of rock had been dumped into the breach, Harriman thought he had finally contained the Colorado. But in October and December of 1906, floodwaters again deluged the area, opening a new thousand-foot-wide gap one-half mile below the recently completed rock dam. Once again, millions of gallons of water began pouring into the Imperial Valley, causing a panic-stricken Harriman to appeal to President Theodore Roosevelt for help. Roosevelt, declining to involve the government in what he considered to be the affairs of a private corporation, responded with the excuse that he could do nothing without the approval of Congress, which had gone home for the Christmas holiday.

Desperate, and spending hundreds of thousands of dollars of his own money, Harriman ordered trainload after trainload of rock dumped into the breach, and in 1907, *two years* after the lower Alamo headgate initially ruptured, the river was finally put back within its own banks.[11] The result of this catastrophe was that thousands of people lost their farms, homes, or businesses, and billions of gallons of diverted Colorado River water had transformed what was the parched Salton Sink into what we know today as the fifty-by-fifteen-mile Salton Sea.[12]

The Imperial Irrigation District

Although the Colorado was back within its banks, calamity had only been temporarily averted. In 1909 massive rainstorms caused the river to flood yet again, threatening a repeat of the 1905-7 devastation. At the same time, after transferring most of its assets to the California Pacific Railroad, the California Development Company filed for bankruptcy. In 1911, a large voting bloc consisting of independent farmers in the Imperial Valley joined together to form the Imperial Irrigation District (IID), which purchased the water rights and diversion materials from Union Pacific Railroad.[13] After continued floods, drought, sediment buildup, swamped levees, and other setbacks, the IID and its boosters beseeched the federal government to do something to protect them from future disasters. Fearing, too, that the Mexican government could cut off their water supply at any time, the IID also clamored for an "All American" canal that would be constructed entirely within US borders. And though the IID—a consortium originally promoted as a democratic organization through which farmers could manage and control their own farmland—was now ostensibly controlling the diversion and distribution of Colorado River water, almost immediately there began a slow and multifaceted wresting of water rights (and thus power and money) away from the farmers.

Ironically, as the political influence of the IID grew, the ability of farmers to determine their own future became increasingly diluted as business owners, developers, land leaseholders, bankers, and lawyers came to control the IID. These groups set about pursuing their own economic and political agendas by slowly replacing independent farmers on the IID board with business and landowner representatives. When the independent farmers realized they were being squeezed out of the decision-making process, many of them complained but were dismissed by the new authority in the IID as dissidents or radicals.[14] Discounting the farmer's complaints as the grievances of a few troublemakers, the new IID was able to safeguard itself from discontented farmers. More importantly, however, the IID was able to accomplish this by also packaging its plans in the seeming neutrality of scientific language and instrumental reasoning so that all of its decisions had the appearance of being so sensible and judicious that ordinary citizens did not feel confident enough to be able to challenge them.[15] As a result, independent farmers had an ever-shrinking influence on the policies affecting them and were quickly dispatched with very little say in the water policy decisions directly affecting their lives.

Although the IID was the most organized, various other small groups of farmer-settlers also attempted to exert political pressure to prod the government into underwriting programs to help family farms and small unorganized farming communities survive in the Imperial Valley area. One such example was Sherman Buck of El Centro, California, who went to Washington, DC, in 1924 to lobby Congress for the All American Canal as an official representative of both the IID and the Imperial Valley Farm Bureau. Writing a letter to his wife from his room at the Hotel Continental in Washington, DC, in March of 1924, Buck expressed his frustration with the delay tactics employed by the secretary of the interior, Hubert Work, writing that Work was "holding up the bill for the sake of delay," and lamenting that affairs were "in a very critical condition." He remarked that if the delays continued, his plan was to go straight to President Coolidge himself: "We are afraid that unless there is pressure brought to bear on [Secretary Work] through the president the bill will be so emasculated that we will not want it. We are thinking very seriously of going direct to the president at once for his intervention. If that is done, and I am urging it, our organization should be represented."[16]

Of course, other, more influential government and corporate entities did not have such troubles eliciting action from the highest positions of power. As outside groups became increasingly omnipresent in determining the political fate of the small farmers, the farmers themselves were becoming increasingly stifled in their attempts to have a voice in these types of decisions.[17] As Donald Worster pointedly describes it, the "evolution of the Imperial Irrigation District is a case study in a familiar twentieth century phenomenon: the collusion between scientifically oriented managers and holders of large private wealth. The first group speaks the language of efficiency, instrumental reason, and environmental domination; the second assents vigorously to that language and proceeds to hire the managers, confident they will end by serving its interest above all others."[18]

Although the IID had at one point become the representative voice of the individual farmer, the land policies of the Imperial Valley were being assimilated by ever-larger political and scientific systems. The discourse and "democratic" process that was seen by the independent farmer as "progress" had evolved well beyond the scope and knowledge of the average farmer. As the move toward large-scale reclamation projects gained steam, the land management practices favored by the US government were planned according to water and irrigation systems and land surveys that spanned thousands of miles and hundreds of political jurisdictions, completely beyond the influence or comprehension of any independent or tenant farmer.

The IID itself was eventually overwhelmed by even wealthier and larger political and business interests: agribusiness owners, the City of Los Angeles, western power utilities, the State of California and other states of the lower Colorado region, the Bureau of Reclamation, the Bureau of Land Management, and the Department of the Interior were all taking up the fight over western water and hydropower.[19] The ability to control the Colorado River brought with it enormous—even hegemonic—political and economic influence in the region, and as a consequence, the small farmer's voice was obscured by the IID, which was in turn obscured by other, more powerful entities.[20]

Nonetheless, as a result of IID and other lobbying efforts, the Reclamation Service, a branch of the Interior Department, issued the Fall-Davis Report in February 1922, proposing that, in order to prevent further disasters, a massive dam be built near Boulder Canyon.[21] In 1922, the Colorado River basin states signed the first Colorado River Compact, and in 1927 the Swing-Johnson Bill, later known as the Boulder Canyon Project Act, was put before Congress. Everything was lining up for the dam to be built—or so it seemed.

Political Obstructions

In the early 1900s, California sought federal backing to build a giant dam on the Colorado River in order to provide drinking water for its booming municipalities and irrigation and flood control for the Imperial Valley.[22] But in order to obtain financing, California needed to come to an interstate treaty agreement with the other states of the Colorado River basin (what would be only the third such instance in US history and the first between more than two states).

The Kincaid Act of 1920 authorized an investigation into the feasibility of a dam and irrigation project for the Colorado River at or near Boulder Canyon. Subsequently, there were two key legislative components to getting the dam project started: the Swing-Johnson Bill and the Colorado River Compact. The Swing-Johnson Bill, put forth by Congressmen Phil D. Swing and Hiram W. Johnson (both of California), outlined the contractual terms of the federal government's role in the dam's construction and apportioned the necessary funding. The compact was needed to reconcile disagreements and prevent interstate litigation among the seven states of the upper (Colorado,

Figure 2.01. "Federal and State Representatives at meeting of Colorado River Commission, Sante Fe, New Mexico," November 24, 1922. *Left to right:* W. S. Norviel, commissioner for Arizona; Arthur P. Davis, director, Reclamation Service; Ottamar Hamels, chief counsel, Reclamation Service; Herbert Hoover, secretary of commerce and chairman of commission; Clarence C. Stetson, executive secretary of commission; L. Ward Bannister, attorney of Colorado; Richard E. Sloan, attorney of Arizona; Edward Clark, commissioner for Nevada; C. P. Squires, commissioner for Nevada; James R. Scrugham, commissioner for Nevada; William F. Mills, former mayor of Denver; R. E. Caldwell, commissioner for Utah; W. F. McClure, commissioner for California; R. J. Meeker, assistant state engineer of Colorado; Stephen B. Davis Jr., commissioner for Wyoming; Charles May, state engineer of New Mexico; Merritt C. Mechem, governor of New Mexico; T. C. Yeager, attorney for Coachella Valley Irrigation District of California. Courtesy of the Boulder City Museum and Historical Association.

New Mexico, Utah, Wyoming) and lower (California, Nevada, Arizona) Colorado River basin, allowing for the river to be developed as one whole unit even though many states claimed individual riparian rights to the water.

Following the Kincaid Act, the seven states in the Colorado River basin set about negotiating an interstate compact whereby the river water would be equitably distributed based on negotiated permanent water rights agreements. The group known as the Colorado River Commission, composed of federal and state representatives, and headed by then secretary of commerce (and future president) Herbert Hoover, met throughout 1922 in an attempt to devise an agreement that was palatable to everyone. On November 24 of that year, at a meeting in Santa Fe, New Mexico, the final draft of the Colorado River Compact was approved (figure 2.01), and legislatures in six of the seven states quickly agreed to its provisions.

Arizona, however, refused to ratify the agreement, citing their concern that, under what is known as the doctrine of prior appropriation (a combination of "beneficial use" and "first come first served," a doctrine fully sustained by the Supreme Court in its *Wyoming* v. *Colorado* decision, June 3, 1922), the ever-increasing demand for irrigation water for Imperial Valley and fresh drinking water for Los Angeles would give those regions a disproportionate share of the water resources, thus preempting Arizona's future ability to develop its own water supplies in the basin.[23] Instead, Arizona demanded a guaranteed permanent allocation of water in exchange for their support, a request the other states refused to grant.

In 1925 the remaining states agreed to a new, six-state compact, but this time California attached a controversial provision stating that the compact's ratification was contingent upon the federal government constructing at or near Boulder Canyon a large water storage and distribution system to make available "sufficient quantities" of water to the lower states (i.e., California). Citing these new demands, the states once again failed to come to terms, and the compact was abandoned.

Nearly two years later, on February 23, 1927, with only nine days left before the end of the congressional session and still with no compact agreement among the states, Senator Johnson found himself in the throes of a round-the-clock filibuster led by the two senators from Arizona, Democrat Henry F. Ashurst and Republican Ralph H. Cameron, and joined by Senators Reed Smoot (R-UT, of the infamous Smoot-Hawley Tariff Act), William King (D-UT), and Lawrence Phipps (R-CO). On February 24 the *New York Times* reported that "for the first time in eight years the Senate, along in the hours approaching daybreak this morning, adopted an order instructing the Sergeant-at-Arms to arrest absentees to obtain a quorum to do business."[24]

As the *Times* described it, "All through the night the battle raged" as members sat in session waiting for a quorum, while the sergeant at arms and his assistants tried to rouse sleeping senators to tell them their presence was required at the Capitol. Senator Harry B. Hawes of Missouri caught a taxicab at two o'clock in the morning, but while en route to the Capitol the cab caught fire. When a fire engine arrived at the scene, Senator Hawes and the cabdriver were reported to be attempting to extinguish the flames themselves. Arthur Capper of Kansas, living alone and not hearing his telephone ringing throughout the night, did not awaken until six o'clock in the morning. When he arrived at the Capitol, he was immediately placed in the presiding officer's chair, relieving an exhausted George Moses of New Hampshire, who had occupied the seat the entire night. Senator Frank Willis of

Ohio, when departing an early morning train at Union Station, was shocked to see the flag flying over the Capitol, indicating the Senate was in session. Immediately telephoning to find out what was going on, he was told by the person answering that he had better get over there in a hurry.[25] Others were similarly confused and sleep deprived upon arriving at the Capitol, and needless to say few were pleased with the turn of events. Opponents of the bill, however, were determined to use any means necessary to stop the legislation, even girding themselves for a legal fight all the way to the Supreme Court.

Meanwhile, in the House of Representatives similar acrimony and confusion stalled the legislation. Representative Elmer Leatherwood of Utah vigorously opposed the bill, contending that "Congress would accomplish nothing by approving the bill in its present form, as it would certainly encounter litigation in the courts."[26] Others offered new interpretations of the bill's intent, or attempted to add unpopular amendments. Representative Earl Michener, a Republican from Michigan, opined that, to him, it was flood control and not drinking water for Los Angeles that was the primary purpose of the bill; meanwhile, Representative Don Colton of Utah proposed an amendment to place power development conditions and contracts directly within the bill, a proposal that was roundly denounced as unworkable given the bitterness between states.

Ultimately, opponents of the bill were victorious, and the bill failed to pass before the March 4 deadline, as the confusion, filibusters, and distrust among senators and congressmen of the basin states were just too much to overcome. After the defeat, Congressman Swing and Senator Johnson set about making adjustments to the bill and formulating plans to discuss and agree upon disputed points with the disgruntled western states—a process that would ultimately take years.

A Contentious Issue for Arizona

The irony here is that the idea of a giant dam to finally control the Colorado River was nearly universally lauded—even by Arizona—as crucial to the development of the Southwest, with politicians, businessmen, and journalists across the country extolling its significance and necessity. Governor Gifford Pinchot of Pennsylvania called it "a great conservation project" that would "add hundreds of millions yearly to the wealth of the people of the United States." President Calvin Coolidge said the project was vital to the national interest. Innumerable others rhapsodized over the dam's potential to transform the southwestern United States. Leo Pasvolsky wrote in the *New York Tribune* in 1922, "The tragic romance of the menace that hangs eternally over the [Imperial] valley has served to focus the attention of the whole nation on a project which may result in the upbuilding of a new economic empire of agriculture and industry within the borders of the United States."[27] Senator Johnson described it as "the greatest constructive project of our time" and a "project of national importance."[28] But in a lengthy 1924 article in the *New York Times* Robert L. Duffus pointed to the primary obstacle in moving forward with the project: Arizona. He lamented, "The lives of 75,000 people and their property in the Imperial Valley, California, are in danger and vast resources in arable land power are kept dormant while long-continued negotiations delay the carrying out of the Colorado River irrigation project." Questioning whether this should continue, he wrote, "The answer depends largely upon voters in one state—Arizona."[29]

The added paradox here is that for years Arizona's politicians and journalists alike spoke of the dam's importance for their state. Nevertheless, they were particularly cautious of two things: the rising political clout of the City of Los Angeles, and the influence of the western power utility lobby. Arizonans felt that, once the federal government was involved, it would overstep its authority and hand to Los Angeles the electricity, water, and taxes that rightfully belonged to Arizona. If Los Angeles city officials, whom the *Coconino Sun* newspaper of Flagstaff, Arizona, called a "gang of exploiters of Arizona," were to secure electric contracts through the federal government, then "they could not under the law sell power to Arizona; they would undoubtedly want tax exemptions because it was a municipal proposition, consequently Arizona would be merely donating its undeveloped wealth to the city of Los Angeles." According to the newspaper, Los Angeles and the federal government would be conspiring to do nothing short of usurping the "birthright" of Arizonans.[30]

Even as far back as the early 1920s, Arizonans were cautioning against any project that would undermine their state's jurisdiction over Colorado River water. In 1922, James B. Girand, at the time a former Arizona state engineer, wrote a long editorial that appeared in the *Bisbee (AZ) Daily Review*—so long, in fact, that it ran in two separate parts on consecutive days, June 28 and 29. In the piece, Girand states that although Arizonans are not in "any immediate danger" of the Swing-Johnson Bill passing, "greedy grasping

California—our dear neighbors into whose coffers we annually pour millions of dollars, and who has no regard for our future growth"—was pressing forward with plans to build a dam on the Colorado River, a project that would, in his words, be "a calamity insofar as Arizona is concerned." Girand was not persuaded by assurances from the director of the Reclamation Service, Arthur P. Davis, that Arizona would receive an adequate allotment of water, dismissing his comments as "meaningless, vapid utterances."[31]

In order to circumvent the federal government's involvement, some in Arizona proposed to build their own dam at Lees Ferry in Coconino County, Arizona, a point that is entirely within Arizona's borders, thus cutting out not only the federal government but also their nemesis, California. An engineer named A. G. McGregor was one outspoken proponent of such a plan. Complaining of the disrespect Arizona was receiving from the Swing-Johnson Bill, McGregor was quoted as saying in 1922 that "if the Swing-Johnson bill were to pass in its present form, this great power resource will be nationalized and Arizona will receive no tax benefit from it."[32] Moreover, McGregor and others insisted that the question of states' rights regarding water had long been settled in previous court decisions giving states prior rights, a position they felt superseded federal involvement.

The argument for states' rights was one that was repeated over and over by Arizonans. Indeed, it was claimed that a federal project to build the dam would amount to an "invasion" and a violation of Arizona's quasi-sovereign rights.

When President Franklin Roosevelt gave his speech at the dam's dedication ceremony in September of 1935, one dignitary who refused to attend was Governor B. B. Mouer of Arizona, who, a few years earlier, sent the Arizona National Guard to "defend" his state against just such an "invasion" by several dozen Bureau of Reclamation engineers. Mouer, like most Arizonans, was highly skeptical that any such dam would safeguard the "entire Southwest," as Roosevelt and others had promised. The Colorado River Compact did not guarantee anything specifically for Arizona, only that 7.5 million acre-feet of water be allotted to the lower basin states. Mouer's worry was that Southern California was growing rapidly, thus requiring ever-increasing amounts of water. Moreover, under the direction of William Mulholland, the City of Los Angeles was pursuing an impossibly aggressive construction schedule for canals and pumping facilities to carry water to the area. Given these realities Mouer was sure that Arizona would never get its fair allotment of water if millions of people in California were already using and depending on it, so he declared that California denying water to a neighboring state in the arid West was tantamount to a declaration of war. So, he began staging a *real war,* one that turned out to be a monumental embarrassment for Arizona.

Aware that the Bureau of Reclamation had already begun test drilling at what was later to become Parker Dam 155 miles downstream from Hoover Dam, Mouer sent Major F. I. Pomeroy and members of the 158th Infantry Regiment of the Arizona National Guard to the border of Arizona and California to monitor the situation and to discourage "any attempt on the part of any person to place any structure on Arizona soil either within the bed of said [Colorado] river or on the shore." When the *Los Angeles Times* learned that the Arizona National Guard had actually been dispatched to the California-Arizona border, they promptly sent their military correspondent to cover the action. With the benefit of paved highways, the *Times* correspondent arrived on the scene well ahead of the Arizona forces because, unfortunately for Arizona, their troops had to traverse a quagmire of sand, rocks, and desert scrub grass, while also crossing several rivers before they actually got to the border.

When Major Pomeroy finally arrived on site, he requisitioned a ferryboat from the town of Parker and promptly renamed his outfit—ironically for a landlocked desert state—the "Arizona Navy." Pomeroy and his detachment tried to take the ferry up the Bill Williams River, but the boat got stuck on a cable and Pomeroy's "navy" was left stranded; however, they were quickly rescued and transported the rest of the way by members of, coincidentally enough, the Los Angeles Department of Water and Power. Once Pomeroy arrived at his final destination, he stayed there for the next *seven months,* radioing daily dispatches to the governor.

One day, as part of the preparations to build Parker Dam, the Bureau of Reclamation began laying a trestle bridge across the river from California toward the Arizona shore. Governor Mouer, asserting that they had no right to build on Arizona land without permission from the state, declared martial law and dispatched a hundred-man militia unit on trucks with mounted machine guns. The standoff between the engineers and the militia persisted until Secretary of the Interior Harold Ickes decided to halt construction until an agreement could be reached. The Department of the Interior then sued Arizona, and in a surprising decision the Supreme Court ruled that

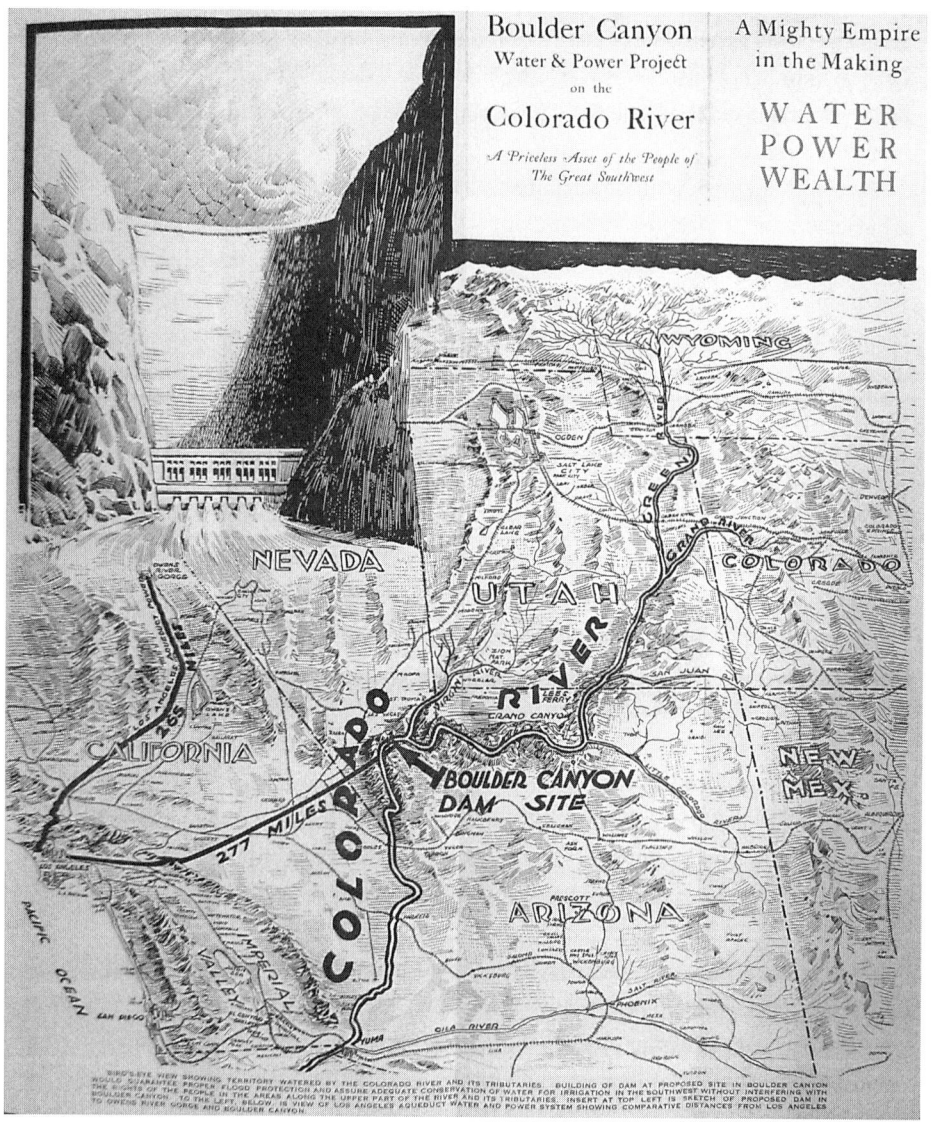

Figure 2.02. "Boulder Canyon Water & Power Project on the Colorado River: A Priceless Asset of the People of the Great Southwest," 1928 (brochure, the Bureau of Power and Light of Los Angeles). Courtesy of Boulder City Museum and Historical Association.

Parker Dam was, in fact, illegal because (unlike with Hoover Dam) Congress had not explicitly approved it. Arizona's victory was short-lived, however, as California's congressional delegation quickly pushed through a bill expressly authorizing Parker Dam, putting an end to the entire imbroglio.[33]

Western Power Utilities

Although the opponents of the Swing-Johnson Bill were suspicious of the earnestness of California's assurance that Arizona would receive its "fair share" of allocated acre-feet of Colorado River water—particularly with the All American Canal and other viaduct projects ready to siphon millions of gallons of water to Los Angeles and Imperial Valley—another concern was whether the development and sale of electricity were going to be controlled by the federal government, controlled by state governments, or left to private business. Arizonans and Californians both were wary of the powerful utility lobby in Washington making a sweetheart deal whereby the federal government used taxpayer money to build the dam but then allowed power companies to siphon all of the profits from the sale of electricity. A lengthy July 14, 1922, article in the *Mohave Country Miner and Our Mineral Wealth*, a newspaper from Kingman, Arizona, presaged much of the concern about the utility companies, stating that all of the "propaganda" in favor of the project was nothing but a red herring cry for states' rights, but in reality it was a power grab by the Southern California Edison company. "It is this company," the article asserts, "that has set afoot all this hue and cry against federal control of the Colorado river and for no other purpose than to grab it off [for] themselves."[34] Senator Johnson himself was also concerned about the political influence of the utilities. Citing the mass of public relations materials put out by the utility lobby, he said in one congressional meeting, "I have a pile of Newspapers, pamphlets, and books two feet high on my desk in San Francisco, all directed against the government's *Hoover Dam* project." "These men," he continued, "seek to stab 60,000 American citizens in the Imperial Valley in the back."[35]

One such example is a 1928 brochure from the Bureau of Power and Light of Los Angeles (figure 2.02), which is illustrative of the hyperbolic publicity materials the utilities were using to sway Americans' sentiments regarding *Hoover Dam*. In this brochure, the Bureau of Power and Light smartly joins grandiose pronouncements about the benefits of the project with an image of the dam dominating a map of the entire southwestern United States.

The rhetorical tactics used in this brochure easily outstrip the often-staid publicity materials of other interested parties at the time. In other contemporaneous publications, images of the dam often portrayed a singular entity alone in a nondescript canyon. More importantly, the images rarely forged any visible connection between the future dam project and the publication's most vital audience: residents and lawmakers in California, and particularly those in the city of Los Angeles. The dam, as characteristically portrayed in other documents, could have been anywhere, while the cautions, benefits, and connections to the project for Los Angelinos were usually only discerned after reading through tedious arguments concerning state and federal regulations, water litigation histories, congressional testimony, landowners in Mexico, power company schemes, and so forth.

The Bureau of Power and Light brochure, however, establishes direct visual connections between the dam and Southern California. The unfurled brochure reveals a giant map of the Southwest with water from the bottom of a towering arched dam flowing out over Nevada and California, feeding directly into the Owens River and Los Angeles Aqueduct. Thick, dark lines connect the city of Los Angeles to both the dam's proposed location and the existing Owens River Gorge and Los Angeles Aqueduct, while the mileage labels suggest that it is virtually the same distance from Los Angeles to the proposed dam site as to the already-tapped Owens River.

Along with imagery connecting the dam project to the city, the pamphlet also uses verbal appeals ranging from nationalist, to populist, to imperialist. Hydroelectric power was obviously the Bureau of Power and Light's main concern, and the headline on the pamphlet's second two-page spread, "Los Angeles Power System Huge Municipal Success," attests to that. The text that follows reminds readers that the Bureau of Power and Light is self-sufficient, "a profitable concern for the city, thoroughly capable of paying its own way," and suggests that a project like Hoover Dam could only serve to make Los Angeles an even bigger industrial power. A section titled "Marvel of the World" reads,

> Los Angeles' industrial development, particularly since the close of the war, is the marvel of the world. Over and over again, in all parts of the world, the question has been asked:
>
> How does Los Angeles do it?
>
> And the answer is:
>
> "Cheap hydro-electric power, supplied by the city at low rates."

In this pamphlet we also see imperialist notions suggested by associating the term "Empire" with the dam project. "A Mighty Empire in the Making" is followed by the words WATER, POWER, WEALTH, all in red, capitalized letters, and then by a call for public support for the Boulder Canyon project "for benefit [sic] of the Entire Southwest."

As empires often are, this was as much a political struggle as it was a technical one: corporatists fought for control of the dam's hydropower; Imperial Valley farmers battled denizens of Los Angeles, Arizona, and the upper basin states for access to irrigation and drinking water; western states were struggling against federal intrusion into what they saw as a states' rights issue; throngs of unemployed men sought to fill precious job openings at the dam site, while unions fought Six Companies and the government over organizing, pay, and working conditions; the Bureau of Reclamation fought against other governmental departments for the right to build and control the dam; and so on. The public relations blitz by the utility companies and other interested parties was a testament to the importance of the Colorado River in the drive to control this new western empire, as access to water meant political power in a region that many at the time viewed as having virtually unlimited economic potential. Meanwhile, in newspapers and other public forums, the state-building rhetoric of empire continued unabated, portraying the river as a useful resource and the dam as both an economic panacea and *the* vital element in developing the American West, an imperative built on an ideology of extending American civilization and using technology to fulfill America's destiny.

Along with its imperialist hue and its usual pronouncements that the project will bring jobs and development to Los Angeles while meeting irrigation demands for Imperial Valley, the Bureau of Power and Light's pamphlet also strikes a populist message by suggesting that this project has "no selfish aims," and that joint ownership and control of the dam between the US government and the people of the Southwest is for the "community good" while still safeguarding the interests of all the communities of the Southwest. The bold headline on one panel reads "The People's Project," while a subtitle claims that the project will be "A priceless Asset to the People of the Great Southwest." The pamphlet goes on to argue that this project is not only economically prudent but in the "people's best interest" and will serve to "combine the various communities and farming districts in a co-operative plan for building Boulder Canyon dam and the development of hydro-electric power." This stance was likely a countermeasure aimed at the skeptical view held by many people of increasingly powerful utility companies and their publicly funded public relations schemes.

I refer to these as "schemes" because there was a sense at the time that public utilities, particularly gas and electric, were involved in elaborate, far-reaching public relations drives (some called it propaganda, others outright lies), at taxpayer expense, to move public opinion favorably in their direction while exerting undue influence in the federal government through their "super lobbies," specifically in the Senate and the US Federal Trade Commission. After the Walsh Resolution, which was supposed to investigate "taxpayer funded public utility propaganda," was defeated in 1928, Ernest Gruening of the *Portland Evening News* wrote an article for the *American Economic Review* that he described as "an attempt to convey a picture of the character and extent of this propaganda." "Just who the originator was of this gigantic plan to indoctrinate the American people," he writes, "seems fairly clear from the record." The article quotes testimony from B. J. Mullaney, director of the Illinois Committee on Public Utility Information, as saying, "Fundamentally the prosperity of our business, its growth and success, are built upon a *proper state of public mind* without which we do not get the money with which to build plants, and we do not get the *favorable reaction from the public mind* that enables us to sell our product to *the best advantage* and at fair rates after we have produced it" (my emphasis). Gruening concluded that the utility companies were producing "the most far-reaching, most elaborate, most expensive, most protean propaganda in the peace-time history of the United States."[36]

Nonetheless, electric utilities such as California Edison were in a strong bargaining position, as the government was caught in something of a catch-22 involving the dam's construction. The utilities wanted ironclad assurances from the government regarding the electricity rates, output, distribution, transmission, and so forth, before they entered into any long-term purchase agreement contracts. However, the government needed the contracts to be in place for lawmakers' review before the legislation could pass. This was a decided advantage for the utilities because they could wait as long as necessary to secure profitable contracts while elected officials endured intense public pressure to pass legislation and get the project under way. The situation stood at an impasse for quite some time, with California Edison at one point threatening to scuttle the entire project, claiming that they could generate electricity more cheaply with their own steam plants than by signing distribution contracts with the government for Hoover Dam hydro-power. Without long-term agreements from utilities, the cost of building the dam would never be recouped, which, considering many lawmakers' antipathy to the estimated costs of construction, would most assuredly seal the demise of the entire project.

A Resolution Without Arizona

On December 21, 1928, H. R. 5773, the Boulder Canyon Project Act, "an Act to provide for the construction of works for the protection and development of the Colorado River Basin, for the approval of the Colorado River Compact, and for other purposes," was finally enacted by Congress and signed by President Calvin Coolidge. However, it was contingent upon a compact being approved by *all* of the upper and lower basin states. By June of 1929 the Colorado River Compact still had not been ratified due to Arizona's steadfast refusal to agree to its terms.[37] But on June 25, 1929, despite Arizona's continued intransigence, Herbert Hoover issued a public proclamation stating that, as a result of the six-state agreement, all provisions of the Boulder Canyon Project Act had been fulfilled and that the project would move forward. Furious Arizona legislators sued the secretary of the interior, Ray Lyman Wilbur, and the states of California, Nevada, Utah, New Mexico, Colorado, and Wyoming, and vowed to take their plea all the way to the Supreme Court, which they did, claiming that "Wilbur is proceeding in violation of the laws of Arizona to invade its quasi sovereign rights . . . and of preventing the beneficial consumptive use in Arizona of the unappropriated water of the river."

By July of 1930 most of the political wrangling (apart from Arizona's, of course), monetary appropriations, and feasibility of the project had been settled or approved. On July 3, 1930, President Herbert Hoover signed a bill allowing for $10,660,000 to be made available to start the project. On July 7, 1930, Ray Lyman Wilbur released an official memorandum for the press stating that "construction of the Boulder Canyon Project had commenced immediately on the President's signature of the Appropriation Bill." In August 1930, the Los Angeles and Salt Lake Railroad Company was contracted to build and operate a railroad from Las Vegas to the construction site at Hoover Dam, while "Construction Camp #1" was erected to house the workers who had already begun to arrive in substantial numbers.

On September 17, 1930, a commemorative ceremony was held at Boulder Junction at which Secretary Wilbur ceremoniously inaugurated construction on the Boulder Canyon Project by driving a silver spike into a railroad tie on the Las Vegas–Hoover Dam line. Photographer Winthrop Davis, who was in attendance, described the unruly scene this way:

> This day was acclaimed a state holiday by Governor Blazer in commemoration of the start of laying the railroad from Bracken, which is seven miles south of Vegas, to Boulder City, 23 miles away by the U.P. Railroad. Secretary of the Interior Ray Lyman Wilbur was to drive a silver spike and give a short address. Wilbur had a rather low esteem of Vegas, and the feeling was mutual with the people of Vegas toward him. The tie had already been drilled to receive the spike, but it was quite evident that Wilbur had never had a spike maul in his hands, for he kept missing it, and the crowd really razzed him. Then came his address, and when he announced that he was changing the name of Boulder Dam to Hoover Dam, the crowd shouted names that Wilbur never knew he had. To top his day off, someone picked his pocket; the rumored amount was four hundred dollars.

As Davis mentions, it was at this ceremony that Wilbur referred to the dam as "Hoover Dam," sparking a controversy over the name that would continue for many years. Originally the dam was going to be built in Boulder Canyon, several miles north of the present dam site. Throughout the 1920s, as the media would report on the Boulder Canyon Project Act, the common reference to the project was "Boulder Dam." After Herbert Hoover won the presidency in 1928, a new bill was put forth in Congress to change the name to Hoover Dam, both in honor of Herbert Hoover's role in the original passing of the Colorado River Compact and for the custom of naming large social projects after influential people. Adding to the confusion was the silver spike ceremony in Nevada where Secretary Wilbur referred to the dam as "Hoover Dam," thus stating a public endorsement of the name change. Although the main appropriations bill, the Boulder Canyon Project Act, retained its name, the project thereafter was regularly referred to and reported on as Hoover Dam. Some of the contracts were rewritten bearing the new name, and new promotional pamphlets and travel guides were sent out referring to the site as Hoover Dam. However, after Franklin Roosevelt defeated Herbert Hoover in the 1932 presidential election, the newly elected administration quickly moved to strike any association of the dam project with Hoover's name. Thirteen days after the new administration took office, the new secretary of the interior, Harold Ickes, issued a memorandum that all references to Hoover Dam were to be changed back to the original name—Boulder Dam. As a result, books, promotional pamphlets, and fliers were again reprinted, this time with the Boulder Dam name. It remained Boulder Dam until 1947, when Congressman Jack Anderson issued a new bill proposing to once again change the name back to Hoover Dam, the third such name change in the dam's short history. The proposition was signed by Harry Truman in 1947, and the name has been Hoover Dam ever since.[38]

ON MARCH 11, 1931, Secretary Wilbur awarded the $48,890,955 contract to construct the dam, power plant, and appurtenant works to Six Companies Inc. of San Francisco, California, a consortium formed by six smaller individual contracting companies who worked together to complete the enormous and difficult project. Six Companies Inc. was allowed seven years from April 20, 1931, to complete the project but was offered incentive bonuses for early completion and finished the project in 1935, almost *two years* ahead of schedule.[39]

On May 8, 1931, the US Supreme Court announced its decision in *State of Arizona v. State of California*, 283 U.S. 423, declaring the Boulder Canyon Project to be constitutional, thus ending Arizona's legal challenges to the project.[40] The majority opinion written by Justice Brandeis concludes that "the United States may perform its functions without conforming to the police regulations of a state," and that "Wilbur is under no obligation to submit the plans and specifications to the state engineer for approval. . . . As we hold that the grant of authority to construct the dam and reservoir is a valid exercise of congressional power, that the Boulder Canyon Project Act does not purport to abridge the right of Arizona. . . . Bill dismissed."

On June 6, 1933, the first concrete batch was poured into a form at the bottom of Black Canyon. Concrete would continue to be poured twenty-four hours a day, 365 days a year for almost two years. On June 16, 1933, at the end of his first hundred days in office and as part of his New Deal economic recovery plan, President Franklin D. Roosevelt signed into law the National Industrial Recovery Act, making available additional funds to complete the dam project through the Public Works Administration. Construction continued at a furious pace until the last batch of concrete was poured on May 19, 1935, and the dam was completed later that year.

3

Illustrating the Dam

Years before actual construction started, people set about articulating through words the dam's importance for the region and visualizing through imagery a conception of a new Southwest aided by a giant dam holding back the Colorado River. In fact, if not for the onset of World War I, there might have been a dam in Boulder Canyon two decades before Hoover Dam. In 1910, businessman Henry C. Schmidt surveyed the area and subsequently applied to the US Department of the Interior and the states of Arizona and Nevada for permission to build a dam and power plant at Boulder Canyon (figure 3.01). Although permits were issued and construction was set to begin in 1914, the project was placed on hiatus upon the outbreak of the war.[1] Although Schmidt's plans were officially canceled in 1922, breathless journalists and fervid politicians continued to expressed their desire for a dam, this one a colossal structure unmatched in engineering history that would bring prosperity—often characterized as an "empire"—to the Southwest. These depictions maintained a consistently promissory hue that situated Hoover Dam as *the* keystone project to ensure the economic future of the region and the linchpin to an even grander imagined future for the Southwest. The project was portrayed both verbally and visually as a panacea that would finally control the threatening Colorado River (a river variously described as "evil," a "menace," "savage," and "unruly") and in doing so provide cheap electric power, water for irrigation and drinking, and protection from ravaging floods for those living in Arizona, Nevada, and California.

Figure 3.01. "Artist's conception of 1913 Schmidt plan for Boulder Canyon dam," 1913 (Nevada Historical Society). Courtesy of the Boulder City Museum and Historical Association.

Illustrations of the dam, particularly in the pre-construction period, ranged from photo/sketch composites, to naturalistic depictions, to wildly imaginative and symbolically rich illustrations, often with a high density of both visual and verbal messages. In this chapter I examine images that, quite literally, illustrated the dam. These illustrations were composed by human hand and were typically drawings that appeared prior to the start of the dam's construction, though some were created during or after construction. I also consider how various illustrations, maps, and composite images—especially those found in newspapers—were used to reinforce an ideology of instrumental reasoning by forecasting what a future dam, and consequently an entire region, might look like.

Object Recognition on a Two-Dimensional Page

Although Southern California and Nevada newspapers published the majority of images and articles on the project, and the earliest illustrations of the Black Canyon area were made as far back as the mid-1850s, one of the first widely circulated visual representations of the dam itself appeared on March 24, 1921, in a most unlikely locale—a small midwestern newspaper, the *Lytton Star* of Lytton, Iowa (figure 3.02). The *Lytton Star* image is important not only because it was one of the first to circulate a visual depiction of the project to the general public, but also because its imagery prefigured three primary features that appear in many subsequent representations of the dam: what I will term *parataxic* depictions of the dam in comparison to another object, superimposing hand-drawn lines and sketches onto naturalistic-photographic material, and labeling the image with numerical data in an attempt to convey the dimensions of the dam. All of these strategies were employed primarily in the service of conveying to a reader the sheer massiveness of the structure, which, when finished, would be the largest

Figure 3.02. "Biggest Dam in the World," *Lytton Star,* March 24, 1921. Courtesy of the State Historical Society of Iowa.

dam and one of the largest civil engineering projects in the world. Furthermore, the significance of this visual-verbal relationship (drawings, captions, headlines, superimposed data, accompanying articles, and so forth) in discourse about the dam is evident from its earliest representations and would remain a crucial feature in subsequent depictions of the project. In chronicling the dam's construction for both engineering and publicity purposes, it is evident that producers of Hoover Dam imagery recognized the challenge of articulating the enormity of the undertaking to a public with no suitable context for appreciating how big the project really was. Thus, they employed several different visual-verbal strategies for those purposes.

There are two key features of the structure that needed to be communicated to the public. First, as was pointed out repeatedly by journalists and statesmen alike, this was at the time the largest civil engineering project in history outside of the Panama Canal. Second, it was a project with a distinct architectural flair, one that changed over time but was just as important to the public's recognition of the dam as were its statistics and feats of engineering. The problem, however, was how to communicate those abstracted ideas (gigantism and design) since, for many of the illustrators, the dam was still years from actually being built.

But before we engage in any in-depth discussion of specific Hoover Dam illustrations, we should first explore how human beings are able to view a two-dimensional sketch and infer from that a real-world mental conception of the object depicted, even if that object is years from physical reality.

In his book *Visual Literacy*, Paul Messaris argues that the development of a three-dimensional perception of a "primal sketch" that illustrates the boundaries and shapes of an object is largely independent from the actual real-world image. In considering approaches to interpreting still images, Messaris suggests several "major discrepancies between concrete-representational images and the appearance of things they represent"[2] and divides his approach into three categories that are helpful for our analysis: light and color, depth, and object recognition. In this chapter I draw on Messaris's categorizations as a framework to address three challenges in representing a yet-to-be-built dam through hand-drawn illustrations: limited strategies for depiction, real-world versus rendered viewing, and feasibility of illustrating size and scope.

First, we can consider light and color in relation to the difference between the real-world visual process of a person seeing the actual dam site or completed dam and viewing a two-dimensional representation of the object. In the real-world viewing of the dam or canyon, aspects of sunlight, shade, and color contrast have a considerable influence on our perception of the object, which neither a black-and-white photograph nor a hand-drawn rendering—the principal means of visual representation at the time—could reproduce. Here we can look to E. H. Gombrich and his groundbreaking *Art and Illusion: A Study in the Psychology of Pictorial Representation,* in which he dissects the conventions of brightness and contrast in pictorial representation, arguing that the *relative* amount of brightness, color, and contrast matters more than the *absolute* amount for human vision.[3] In other words, a representation cannot reproduce, for example, the brightest or darkest aspects of the real-world object or image; however, a sense of brightness and darkness can be conveyed through contrast internal to the representation itself such that a viewer can interpret that representation through mental skills already developed by interacting with other images and the outside world. Thus, the exact reproduction of color and light are irrelevant for what Aristotle called the main purpose of vision: figuring *what* is there, and *where* it is.[4]

Another constraint to conveying information through hand-drawn representations of the dam was certainly the prosaic medium of early 1900s newspapers, in which viewers of pre-construction Hoover Dam images were looking at black-and-white, hand-drawn renderings consisting of almost no naturalistic light, color, or contrast. Even in superimposed sketches over photographic images, the hand-drawn renderings were simply outlines of shapes against dark, grainy pictures of canyon walls and river water. Messaris argues, however, that the development of a three-dimensional perception of a "primal sketch" that illustrates the boundaries and shapes of the object is largely independent from the actual real-world image, and that there is no evidence the absence of naturalistic color or light will necessarily prevent a pictorially inexperienced viewer (a viewer who has not seen the real-world manifestation of the thing being represented) from being able to comprehend the content of an image.[5] In other words, simple line drawings or black-and-white photographs lacking naturalistic light, color, or contrast do not require any special interpretational skills or experiential association with the real-world image on the part of the viewer in order for them to ascertain the image's content.

But, regardless of whether or not a viewer can decipher the content of an image based on rudimentary outlines devoid of naturalistic colors or light, it does not necessarily hold that an image is able to impart or even suggest the size or scope of a real-world object. Therefore, we should consider a second characteristic in the relationship between representational images and the object itself, which is *depth*. Depth, or stereopsis, is the awareness of the relative distance between an observer's eyes and some point in the visual field.[6] There are several components of depth in theories about real-world vision, such as binocular retinal disparity, motion parallax, occlusion, texture gradients, contours, shading, and so forth; nevertheless, not all of these components are needed for viewing still images, and only one is important for our analysis here: *contours*.

The outlines and contours of an object allow the brain to see it as three-dimensional, but according to prevailing theories of vision, the inference of depth precedes the identification of the actual object. In other words, the brain sees the solid objects in three-dimensional precept before their identity has been established. Thus, the brain knows, for example, that an object is a dog several feet away before it knows that the dog is your pet Fido. Therefore, when images representing humans, animals, and other forms are shown through outlines and contours, the strategies that our brains use to interpret the world through three-dimensional depth perception can be brought to bear on aspects of two-dimensional imagery in which these particular forms occur.[7] The perceived depth, or distance, between the viewer's eyes and the representation of something like Hoover Dam on the page is obviously going to be much different from that of a viewer standing on the precipice of a canyon. Consequently, the artist creates contours in the image sufficient to produce a three-dimensional precept that can then be identified as a large object between two canyon walls. But this would seem to only be sufficient to produce a three-dimensional, identifiable image—nothing approaching a faithful representation of the object, or even one approximating the dimensional properties of the object.

To address this issue, we can consider another aspect of spatial representation and depth cues: perspective (also referred to as "Renaissance Perspective"). It is possible that an image has depth as a result of contours with which the viewer is familiar, or contains shading and brightness (particularly in the case of photographs) such that the changes to the surface of the object relative to the light source produce depth cues; however, perspective invites us to relate gradually smaller features in an image with increasing distance.

The characteristics of depth and dimensional representation we are typically most familiar with are *linear perspective* and *relative size*. Linear perspective is the idea that objects receding away from the viewer follow parallel lines in the picture plane that eventually coalesce into a "vanishing point," creating the illusion of depth and distance on a flat surface. Although the vanishing point has been used in art since at least the early 1400s and is certainly present in illustrations of Hoover Dam, the key to our analysis is the notion of relative size, what Messaris describes as an "inverse relationship between depth and the size in which objects are portrayed." Because the aim of most early Hoover Dam imagery was to express to readers the sheer magnitude of the object, relative size was an essential element that allowed artists to use what I call a "parataxic" visual relationship (*parataxis*, παράταξις from the Greek *para* meaning "beside" and *tassein* meaning "to arrange," which together might roughly translate to "act of arranging side by side") of images in which familiar landmarks and other comparative objects are superimposed against photographs or illustrations of the dam, a strategy that we will see used repeatedly in representations of Hoover Dam. The premise of this strategy is simply that if a viewer of an image is at all familiar with the "real-world" size of a particular object, then the size of that object portrayed in a picture and operating as a depth cue can serve as comparative information about the relative depth and size of other things depicted.[8]

Although used extensively in Hoover Dam imagery, this strategy is not unique to the dam, as such conventions had been used with relative frequency before. The full front page of the December 5, 1909, edition of the *New York Tribune* is just one example in which pyramided visualizations of the amount of refined sugar, tobacco, coffee, and raw sugar consumed in the United States in 1908 are compared to the Washington Monument, the New York municipal office building, the Pyramid of Cheops, and the New York Tribune Building.

Although the seeming congruity of using these kinds of parataxic visual strategies of relative size appears to explain how viewers are able to get a sense of real-world size through hand-drawn illustrations with little color, contour, or contrast fidelity, Messaris goes on to argue that relative size is "unlikely to be a real-world depth cue in the normal sense, because depth perception occurs before object identity (and the meaning of relative size) has been established."[9] However, he makes this point by citing two influential studies in which people shown at varying distances in distorted or plain backgrounded images were perceived only as having different *sizes*, not different depths or distances. Thus, this does not impugn the use of relative size to explain dramatically spatialized representations of the dam and, in fact, supports the argument that relative size can provide viewers a sense of scale using comparative imagery. There remains, however, one additional condition of our interpretational and visual processes that must be met in order for relative size to work, and it is one that I've already mentioned: *object recognition.*

Contemporary notions of visual object recognition suggest that even though pictures may be vaguely structured or vary tremendously in their surface appearance from a real-world object, minimal implicit structural information (even a composite or "bad" representation) is all that is required to satisfy the brain's object interpretation and recognition processes. Consequently, hand-drawn, two-dimensional images—even simple or "bad" ones—can satisfy our visual recognition requirements without actually replicating many of the viewable features of the real-world object.[10] This is clearly evidenced from our ability to recognize and render stick figures that represent mommy, daddy, and Fido, or structures such as houses, all at a very young age.

This necessarily brief review allows us to establish that several assumptions of human visual processes are *not* germane to a person's ability to interpret a two-dimensional representational image of an object: naturalistic light and color; incongruences between perceived depth in the viewer/real-world-object and viewer/two-dimensional-representation relationships; and "realistic" or explicit, detailed structural reproduction of the object. Therefore, we can say that to convey a sense of size and scope of the dam on a two-dimensional page, someone need only illustrate minimal structural information using *contours* and *contrast* such that a viewer can interpret the *relative size* of the object as compared to other *recognized objects* or ideas with which the viewer is already familiar.

Although this notion of visuality makes it possible for a viewer to grasp the content and relative size of a represented object, there is a vitally important aspect missing here, which is the fact that nearly all of the two-dimensional illustrated representations of the dam use contours, contrast, relative size, and recognized objects, but they also use *verbal text*. Whether through data markings that include length, width, height, weight, or mass or effusive adjectives to describe the object, text is the adjoining fundamental feature of the illustrated renderings of the dam that image producers employed to articulate verbally what is very difficult to do pictorially.

Early Reliance on Verbal Messaging

Although portrayals of the dam would later move toward more reliance on visual imagery to tell the story, early descriptions of the pre-construction dam were heavily influenced by verbal messages. The first thing one notices when looking at the *Lytton Star* image is the large, bold headline that reads "Biggest Dam in the World." To help viewers get a sense of how high the dam will be when completed, the authors depict a parataxic relationship between the dam and the Washington Monument. We see in that image the dam represented by an area of shading between the canyon walls with a vertical line labeled "286 ft" and a horizontal line labeled "1080 ft," along with a caption that reads "Colorado Dam Compared with the Roosevelt Dam." In the middle of the shaded area we see a drawing of the Washington Monument labeled as such and accompanied by another label indicating a height of "555 ft." Above the Roosevelt Dam and the Washington Monument is a dotted rectangular box with a height label of "572 ft," which was the projected height of the dam to be built on the Colorado River. If the illustrator of this image had the benefit of additional information, the discrepancy could have been shown even more dramatically, as the actual height of the completed dam is 726 feet. Here, though, we see an attempt to express the magnitude of the dam through drawn contours and shading, relative size of the object as compared to other recognized objects, and labeling through the headline as well as textual data markings of scale.[11] Although the *Lytton Star* had negligible circulation numbers compared to other newspapers, this image of the dam presaged a much wider distribution of similarly structured imaginative renderings that had greater public impact, particularly in the Southwest.

Published four months later, an article in the *Los Angeles Examiner* echoed the projective imagery seen in the *Lytton Star*, although labels in the *Examiner* indicate figures somewhat closer to its actual finished height (figure 3.03). The *Examiner's* image has hand-drawn features over a grainy photograph of a canyon with lines that form a T with the vertical labeled "600 ft" and the horizontal labeled "Top of the Dam." As in the *Lytton Star*, this image attempts to convey the tremendous height of the dam; however, its abstracted label of "600 ft," without comparison to a familiar object providing a sense of scale, has no hope of conveying the real-world size of the dam to a reader. Instead, this representation relies almost entirely on verbal text to communicate its scale. The headline "Mammoth U.S. Dam at Boulder Canyon Promises Vast Wealth and Unrivaled Power for Los Angeles" does the work of coalescing the notions of prosperity and power with an imagined projection of the dam's dimensions in order to evoke the significance of the project.

The *Lytton Star* and *Los Angeles Examiner* pages do share a number of other similar design features. We see in each a headline at the top of the page that conveys the project's significance, while composite photographs/renderings are placed underneath for purposes of illustration. Though more prominent in the *Lytton Star*, both have ornate, art nouveau–like decorative features.

Figure 3.03. "Mammoth U.S. Dam at Boulder Canyon Promises Vast Wealth and Unrivaled Electric Power for Los Angeles," *Los Angeles Examiner*, July 24, 1921. Courtesy of the Boulder City Museum and Historical Association.

The *Star* includes illustrated foliage around the composite image, while the *Examiner* includes two hand-drawn corbels at the bottom of its composite. Both have several panels, each containing a photograph, a composite, or a map. (The map is quite conventional in a news story of this sort and is a particularly significant feature in later depictions of the project, a topic I discuss later in this chapter.) Each also includes a separate photograph of the canyon in addition to the illustrated rendering. The *Lytton Star* includes a prominent facial portrait labeled "A. P. Davis" (Arthur P. Davis), director and chief engineer of the Bureau of Reclamation who initially proposed building a dam of "unprecedented height" in Boulder Canyon.[12] The *Examiner*, too, presents a facial portrait, this one of E. F. Scattergood, chief electrical engineer of the Los Angeles Department of Water and Power. A major difference, however, is in the visual-verbal relationship of each.

Davis's portrait stands out on the page with a white background and circular frame set against the dark canyon wall of the photograph. The associated text, however, is simply a label of "A. P. Davis" with no other contextual information. It is not until roughly halfway through the article that Davis is mentioned. The article then proceeds to quote him for several paragraphs, in which he describes the Colorado River as "unruly" and a "menace" (terms we will see repeated countless times in subsequent descriptions). Conversely, Scattergood is shown less prominently than Davis—Scattergood's image is lower on the page and incorporated into a smaller article. However, the Scattergood portrait is accompanied by the promissory headline "Great Benefits of Power Urged by Scattergood."[13] This is indicative of a broader difference in the two newspaper presentations. The *Star*'s main headline is effusive, but it seems rather pedestrian when compared to the ostentatious description in the *Examiner*. While the *Star*'s headline is only about the dam itself, the *Examiner*'s headlines "Promise" to all of Los Angeles "Great Benefits," "Vast Wealth," and "Unrivaled Electric Power"—a future for both the dam and region far grander than that intimated in the *Star*.

Another article, this one in the January 15, 1932, *Progress Edition* of the *Las Vegas Evening Review-Journal* and written by then commissioner of reclamation Elwood Mead, explains the extraordinary changes taking place at the dam site. The article is accompanied by yet another photograph of the canyon with superimposed hand-drawn dotted lines and text outlining future aspects of the project: the top of the dam and its elevation; locations of the diversion tunnels; roads for trucks; a lookout point and its elevation;

footbridges across the canyon; power lines; and so forth (figure 3.04). The article's title, "Mental Tour over Nation's Greatest Job Astonishes," provides a pretext to Mead's portrayal of the project as a sublime undertaking, one that, in order to measure its progress, requires that we look back to the previous year when "the site of Hoover Dam and Boulder City stood in all the stark nakedness of the desolate and forbidding desert. The visitor today is amazed at the astounding change which has been wrought in a few short months. On every side he sees the tangible evidences of man's conquest of the handicaps imposed by Nature." It is important to note here, too, that Mead's description of the project as one in which man is conquering the "handicaps" imposed by nature presages what will become the dominant trope in future descriptions of the dam and its relationship to the Colorado River and surrounding landscape.

Figure 3.04. "Mental Tour over Nation's Greatest Job Astonishes," *Las Vegas Review-Journal*, January 15, 1932. Courtesy of the Boulder City Museum and Historical Association.

Another headline, this one in the June 16, 1933, edition of the *Las Vegas Evening Review-Journal*, reads "Stage Is Set for World's Greatest Dam: Graphic Picture Story of Man's Triumph over Mighty River." Although the huge image on the page is similar to other graphic illustrations of the construction site, this one is extraordinarily detailed. In a nearly three-dimensional view, the artist depicts the cableways stretched across the canyon lowering men and materials down to the riverbed; outlines of the technical drawings and specifications for the diversion tunnels, height of the dam, and outlets on the Arizona and Nevada sides; observation points along the canyon rim; "himix" and "lowmix" concrete mixing plants; the highway to Las Vegas; Boulder City; future roads and electricity lines; and more. These examples suggest that, for this stage of the project, the editors and publishers of these newspapers prejudiced the impact of grandiloquent words over the ability of images to sway public opinion.

In his introductory chapter to *Ways of Seeing*, John Berger articulates a set of concerns with images (photographic, painted, or drawn) and their relation to text by claiming for the image a prior and more central place in the human sensorium, arguing that it is "seeing which establishes our place in the surrounding world; we explain that world with words, but words can never undo the fact that we are surrounded by it."[14] For him, words are a reduction of the image, an attempt to capture through language the essence of the visual, which will inevitably elude that attempt. However, in these examples we can see that the opposite is true. The images are a reduction of the word, subservient to the headline and its ability to quickly and powerfully articulate meaning to a reader. This seems to be especially pertinent for those attempting to depict or represent the seemingly incomprehensible magnitude, or the sublimity, of the project. Though I cover the sublime and Hoover Dam imagery at length in the next chapter, it is worth noting here that in the case of Hoover Dam, the sublimity of the structure—with its sheer enormity, technological splendor, and ability to tame the untamable forces of nature—is an unmistakable visual and particularly verbal hallmark of many of these illustrated representations. The sublimity of the dam is something that image producers struggled to visually articulate, and, realizing the limits of illustrating a sublime object on the printed page, they often resorted to grandiloquent words to convey their meaning.

Although the enormity of the project, particularly in pre-construction illustrated representations, is communicated principally through words supported by images, we will see in later representations a distinct shift away from the simplistic depictions in these examples to increasingly sophisticated, visually complex, and eventually symbolically charged imagery that is more capable of conveying the majesty of the undertaking.

Architectural Design of the Dam

The most visibly obvious physical characteristics of Hoover Dam relate to its architectural design, an instantly recognizable and aesthetically pleasing art deco style that thrust the dam into the public consciousness as an icon of modernism. In an early newspaper article that carried images of the dam, the September 3, 1929, *Jefferson City (MO) Post-Tribune* includes on page 9 the usual awe-inspiring headline "3000 Men Toil for Seven Years in Building of Huge Boulder Dam," yet in that article we also find a fully developed architectural vision of the completed dam without the sketched lines and statistical labels typical of other contemporaneous depictions. This article features two images: "Today," a grainy photo of the canyon without added detail, and another image below showing a comprehensive rendering of what engineers imagined the completed dam would look like in 1937 (figure 3.05). Again, the importance of the verbal text is evident in that the headline, not the image, suggests the immense effort required of laborers to finish the project, while the "Today" and "In 1937" labels convey the temporal relationship between the images that we can only understand through verbal text. Yet by showing the dam façade, power plant, and appurtenant works in a more detailed rendering than had appeared previously in newspapers, this article also provides another aspect of meaning that the text cannot: a sense of design.

In the preliminary engineering sketch we can see architectural and design elements such as the large eagles atop the dam and the decidedly Greek structural influences for which the original drawings were roundly panned. To counter these criticisms, the bureau sought outside input and hired architect Gordon B. Kaufmann to salvage a design that some commented was reminiscent of Hugh Ferriss's futuristic "Metropolis of Tomorrow" drawings and similarly stylized scenes often found in comic books.[15]

From December 1930 to November 1932, Gordon Kaufmann, a Southern California–based architect then best known for his design of the Los Angeles Times Building, made recommendations to the Bureau of Reclamation regarding the design of the dam,[16] and in January of 1933 he presented the

Figure 3.05. Detail of preliminary architectural sketch "In 1937." Courtesy of the Nevada State Museum.

bureau with a preliminary sketch including various changes to the original design that were, for the most part, implemented.[17] In his redesign, Kaufmann simplified and modernized various aspects of the dam, such as replacing the overhanging balcony and imbalanced towers with a series of observation niches and towers that rise upward unobstructed.[18] Kaufmann also added lighting to the top of the intake towers and recast the powerhouse in a more modernist-classicist style.[19] Other aspects of the redesign include the fenestration of both the power plant housings and the façade of the needle valve housings. These were places where the engineering requirements to endure the physical demands of a working dam came together in a synergy of form and function with architectural choices such as the use of concrete cast against plywood to provide as smooth a surface as possible.

Though it often goes unnoticed, the color of the dam itself was another consideration for Kaufmann. The entire dam structure is made of layers of poured concrete, and so controlling for color variations in the numerous individual mixes posed a challenge. Variations in the quality of the curing water used in each separate mix, combined with the different types of pipe conducting the water, would result in a lighter or darker cured concrete color depending on the water and pipe combination used. With the ability to affect the cured color by changing water quality or piping materials, Kaufmann, in consultation with artist Allen Tupper True, made the decision to subtly grade the concrete color from darker at the bottom to lighter at the top.[20]

In an article outlining his approach to the design, Kaufmann wrote that there was never an attempt to create architectural effects or features that would not fit into the design of the dam structure as a whole. Although beautiful aesthetics are typically not the first consideration in projects such as this, with Hoover Dam the structural design and architectural design were complimentary features, with neither one dominating the other. The curved shape of the dam and location of the elevator towers, for example, were determined by the original engineering schematics. Kaufmann placed the utility towers in reference to the elevator towers so as to create what he described as a "rhythm" in the overall design. The viewing platforms were placed with reference to the horizontal joint lines that cross the face of the dam, "forming an orderly series of small vertical shadows punctuated by the larger shadows of the elevator and utility towers." The only true examples of "ornamentation" on the dam, according to Kaufmann, are the monolithic concrete sculptured panels created by artist Oskar J. Hansen, which emphasize the entrance to the passenger elevators and represent, on one panel, irrigation, navigation, power development, flood control, and water supply, while the seals of the surrounding seven states adorn the other.[21]

All of these design changes were integral to the dam becoming a symbol of modernity—it visually radiates the tenets of modernism through its design. As a work of modernist architecture, the dam is emblematic of several key themes of American modernism: efficiency, advanced technology, ahistoricity, and an emphasis on simplicity and utility.[22] Kaufmann wrote that the dam surely is a wonder of the world with its massive size and its technological execution, and that it will remain so until such a time as it is eclipsed in height, scope, and concept by other dams. But to him the dam was more than its superlatives—it was "architecture on a grand scale."

The architectural style is important because it is not only a defining feature of the visual iconicity of the dam but also representative of a wider
transition in American thought, values, and culture. Jean Baudrillard argues
that, in contrast to the Europeans, who were caught up in the ideological
and revolutionary fervor of the nineteenth century (as evidenced in the
French Revolution), the Americans remained aligned with the utopian and
pragmatic philosophies and moral perspective of the men of the eighteenth
century. For Americans, it was the architectural styles of the eighteenth and
nineteenth centuries that came to epitomize the excesses and vulgarity of
the arts.[23] Consequently, the modernist response to this was a return to simplicity and purity.[24]

Blair, Jeppeson, and Pucci argue that modernist architecture wants to
be "purged of every historical or symbolic contamination," and that rather
than being an art form in itself, the aim of modernist architecture was to
"signify the twentieth century's achievements and dominance of technological innovation, rationality, and corporate power."[25] But it was not just
witnessing the design in the physical presence of the dam that achieved this.
Through its dissemination in imagery of various kinds, the architecture of
Hoover Dam became a key feature of its development into a symbol of modernism by visually projecting its ethos of ahistoricity, efficiency, and technological control not just across the canyon in which it is situated, but to the
entire nation.

Ideologies of Conquest in the "Los Angeles Examiner"

Although we have already seen some examples of the promotional and celebratory accounts in newspapers highlighting the exalted nature of the
dam and its importance for the future of the Southwest, none can rival the
extravagant and unrestrained celebration of the dam project shown in the
July 16, 1929, edition of the *Los Angeles Examiner* (figure 3.06). The *Examiner*,
originally a union-friendly paper whose motto "A Paper for People Who
Think" graced the top of each page, debuted in 1903 as part of William Randolph Hearst's campaign to secure the Democratic nomination for president. Although the paper shifted to decidedly more conservative views as
time went on, eventually becoming aggressively opposed to the labor movement and immigrants, and particularly to Mexican Americans, the first edition of the *Examiner* began with the headline "All Southwest Welcomes the
Examiner; City Never Had Such a Demonstration," followed by the subtitles

"Streets of Los Angeles Crowded with Marching Citizens Cheering for William Randolph Hearst," and "More than Ten Thousand Members of Labor Organizations and Unions Salute *Examiner* Office," and finally "Hearst for the White House." The *Examiner*'s chief rival, the *Los Angeles Times*, however, had a different take on the public's enthusiasm for the new daily. Their editorial, which appeared on the same day, read, "A very careful count . . . gave 2,843 the exact number of marchers . . . somewhat less than the 11,000 expected, and the unionists were somewhat dissatisfied with the turnout . . . when the parade passed the new office of the *Los Angeles Examiner* there was a little weak cheering by one or two unions in the line."

The full-page illustration in the July 16, 1929, *Examiner* is particularly striking for its size and layout. At the top of the *Examiner*'s "Boulder Dam and Expansion" edition front page, two text boxes indicate to the reader that this is a special issue for a momentous project. The top left reads "Boulder Dam—One of the greatest achievements of all history," while the right reads "Mail this edition to your friends and relatives in the East." In the center of the page we see another block of text, this one with typewriter font to indicate that the source is something other than the newspaper's editorial. The text is the official "Public Proclamation" issued by President Herbert Hoover and Secretary of State Henry Stinsman that the provisions of the Boulder Canyon Project Act, originally approved on December 21, 1928, had been met. This proclamation announced that California, Colorado, Nevada, New Mexico, Utah, and Wyoming had ratified the compact and waived the provisions of the first article of the Boulder Canyon Project Act, which required that all seven states (including Arizona) agree to a negotiated compact.[26]

On the *Examiner*'s front page we see an imaginatively complex illustration of the dam, this one a massive wall extending straight across the canyon. Where previous depictions had presented a plain, arched structure, a conception evidently derived from initial designs issued by project engineers, this version is almost castle-like. With its turrets and parapets at the top and arched loops across the middle, it is a more forceful, militaristic presentation than the smooth, concave design featured in earlier images.[27]

This illustration is reflective of the ideologies inherent in much of Hoover Dam imagery that invoked political and economic interests, territorial acquisition, and notions of empire that often took on an imperialistic and overtly militaristic hue portraying nature as an enemy that needed to be conquered. The images that depict man's technological subjugation

Figure 3.06. "Boulder Dam and Expansion," *Los Angeles Examiner,* July 16, 1929. Courtesy of the Boulder City Museum and Historical Association.

of the natural world are, however, often wrapped in aesthetically beautiful or awe-inspiring visual presentations: the primary mechanism through
which ideologies about the dam's role in man's relationship to nature are
communicated.

Ideologies are the pervasive, shared, yet often mundane and taken-for-
granted values, beliefs, and assumptions that permit an individual to function within the complex relationships and activities of a group of people. But
the reason they are powerful is that they are often deeply embedded in the
quotidian: the diurnal endeavors of everyday life such as reading a newspaper, or viewing a brochure.[28] Images of Hoover Dam echo assumptions
and attitudes that were characteristic of modernist ideals, which were made
to seem normal, natural, or given, as opposed to being part of a culturally
constructed belief system. The ideologies projected by the images of Hoover
Dam often show the dam process as an inevitable, even natural, progression
of remaking the Colorado River to the way that it should be: controlled by
the Bureau of Reclamation and distributed equitably throughout the region.
These cultural beliefs and attitudes about the "untamable" or "unruly" river,
the inhospitable wasteland of the desert, and the squandered water that
flowed unused into Mexico and then to the ocean were both constructed
and lived experiences recorded by artists and photographers and distributed to the public through various media outlets. Thus, the visual rhetoric of
Hoover Dam imagery not only reflects the ideologies and power structures
of the Bureau of Reclamation, business interests, and so forth, but is fundamental to them.

The dominant feature of the *Examiner* illustration is the large, brawny
man standing in the middle of the image looking at scrolled paper, possi-
bly a survey map or schematic drawing. The man's sheer size and placement
at the foreground of the image command our attention. He stands with a
relaxed pose, weight leaning on one leg. His expression is that of someone in
calm contemplation, but his is also the only face that is discernible, as those
of the other human figures appearing in the scene are either undefined or
hidden in shadow. The man's clothing consists of boots and field pants, a button-down shirt complete with cigarettes in his pocket, and a hat seemingly
rumpled by years in the field. He is also a symbol of strength, confidence,
intelligence, and leadership, a man with the musculature and clothing of
the working-class everyman, yet one who is reading schematics, not toiling
and covered in dirt like the others shown behind him. And, unlike the other

figures in the image, he is not soiled by the machinations of industry. He is not drilling rock or scooping dirt amidst the plumes of smoke rising from the construction activity; instead, he appears contemplative, cerebral, and commanding.

In this illustration we see all the workings of industry: machines, trucks, laborers, exhaust, scaffolding, cranes, electric towers, and power lines. One thing we do not see, however, is water. We can surmise that this is the Boulder Dam construction site from the headline and the large structure in the background, but the Colorado River, the project's raison d'être, is conspicuously absent. A primary motivation for this project, as we have seen several times now, was flood protection from the "black menace" that was the Colorado River, but that threat is not depicted or even suggested; instead, the threat has already been neutralized, the river subdued and driven back behind the massive walls of the dam.

The image is also highly militaristic. The figure on the left drilling the rocks brings to mind the machine gunners of World War I. He "fires" into the wall and pushes the enemy back, keeping it pinned down so that the other men and machines can continue to work free from immediate threat. The man in the forefront also evokes the sense of a field general examining his battle plan as the soldiers work to build the dam and dredge the soil. The men and machines appear to be at war with nature. As the chaos of construction continues below, the dam itself—a fortress against nature—looms in the background.

This image echoes the visualized hypermasculinity that appears in images throughout the construction of the dam, particularly in the continued use of military images and metaphors similar to the "Boulder Dam and Expansion" edition of the *Los Angeles Examiner*. In these portrayals, militarized depictions in both word and image frame the Bureau of Reclamation's task as a pitched battle to control the Colorado River.

In the March 24, 1931, edition of the *Las Vegas Review-Journal* an article headline reads "Hoover Dam Contractors Mobilize Army for 6-Year Task of Harnessing Colorado." The article, written by Frank Frawley of the Associated Press, begins, "After more than eight years of planning and agitation, the major offensive against the turbulent Colorado River has been started." An article in the August 1932 edition of *Construction Methods* magazine carries the headline "Feeding the Hoover Dam Army" and includes several images that evoke military-style mess halls. The article states, "Under the direction of Six Companies Inc., contractors for building Hoover Dam, the Anderson Bros. Supply Co., of Los Angeles, is keeping the Hoover Dam army well fed. The charge is 50c. per meal." The October 1934 edition of *American Legion Monthly* includes an article by Alexander Gardiner titled "Out of the Desert—an Empire," playing on a particularly martial theme and yet another reference to the Hoover Dam as part of a new empire in the Southwest. This article includes an image of a "drilling jumbo" that was reprinted from a 1932 Bureau of Reclamation photograph (figure 3.07). The caption describes the men in the image as "machine gunners of peace." Similar to the image in the *Examiner* discussed earlier, the drilling mechanisms evoke the machine guns of World War I, and the appearance of the men in the photo further solidifies the military allusion. The men are all dressed in coverall uniforms, each wearing a hard hat similar to soldiers' helmets. Each of the foremen

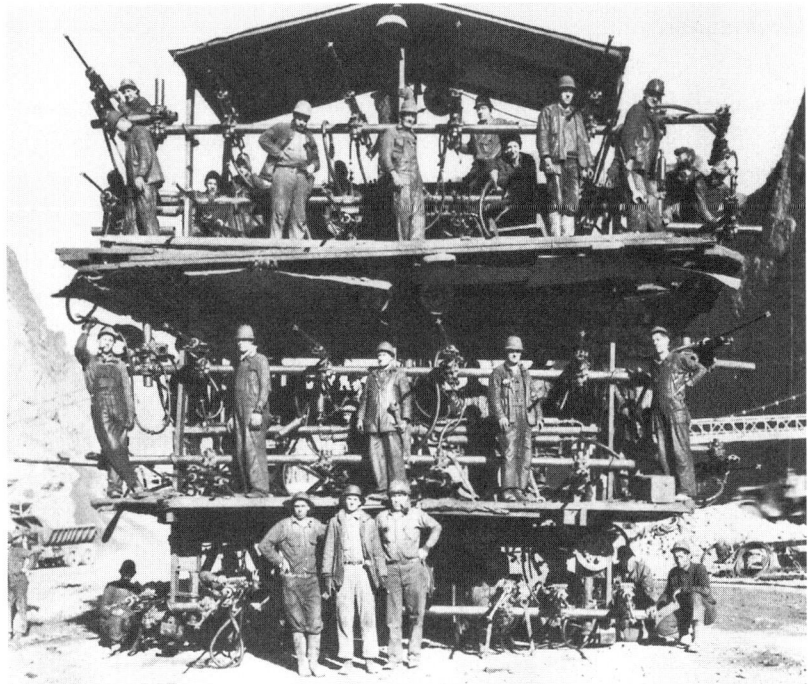

Figure 3.07. "30 Head Drill Jumbo," April 4, 1932 (photograph, Bureau of Reclamation, later reprinted in *American Legion Monthly,* December 1934). Courtesy of the US National Archives and Records Administration.

standing on the ground to the right of the platforms has on a hard hat, field jacket, and pants similar to a military commander in the field. These men, with their "machine guns," would be wheeled up to the canyon wall to drill through yards of solid rock.

The striking front-page image of the *Examiner* is a grandly symbolic exhibition of man's dominance over nature and the taming of the once-feared river. The visual and verbal magniloquence is just as intense on subsequent pages of the "Boulder Dam and Expansion" edition, where nearly every article is devoted to the dam, each one trumpeting the economic benefits of the project, with headlines reading "Cheap Power Guarantees Huge Growth," "Upper Basin Areas to Profit," "Project Means 12 Billions New Wealth for Southland," "Entire Southwest Will Reap Enormous Gains," "Long Prosperity Era Predicted by Monnette," "Trade Outlets Here Will Be Quadrupled," "Mead Visions Great Center of Industry," "4,000,000 Expended During Next Decade," and "State Leaders Tell of Boulder Dam Benefits."

The *Examiner*'s editorial images and articles were not the only promotional materials in the edition. Real estate, gasoline, and concrete companies; the City of Needles; California's Chamber of Commerce; Pacific Electric Railway; the City of Los Angeles's Water and Electric Systems; and even the paper itself ran advertisements that used images of the dam and sought to forge visual and verbal linkages between their interests and the optimistic projections for the project. For example, a real estate promotion for Le Clair Rancho boasts a "green paradise of productive crops" and "Fruit, Cotton, Dates, Etc., Plenty of Water!" To highlight the abundance of water in the middle of the Mohave Desert, the advertisement also features a photograph of a man in a scrub-grass landscape standing next to a gushing well, evidently in Le Clair Rancho. An ad for Rio Grande Gasoline and Lubricants asserts no explicit verbal connection to Hoover Dam, but features a large rendering of the dam with the word "POWER!" in uppercase, bold, art deco lettering below it. The City of Needles, California, lays claim in another advertisement to being the "most strategically located city in the lower Colorado River Basin" at the "Center of the Boulder Dam Empire," a "Metropolis of Eastern California" that will "undoubtedly be one of the largest beneficiaries of the Colorado River development."

Next to the Needles, California, editorial is a page-length advertisement from the *Los Angeles Examiner* itself. The ad features a highly stylized, impossibly narrow dam spanning two relatively smooth canyon walls with the word "Achievement" serving as its base. In a small box below, a man sits on horseback against a backdrop of a mountain and clouds. The ad features art deco styling in both the dam and the text, while the ad copy provides insight into why the newspaper ran such an edition in the first place. It states,

Public opinion is building Boulder Dam!
When completed it will be a monument to the will of the people; a triumphant achievement, the result of a ten-year fight against the most powerful opposition. The Los Angeles Examiner is proud of the fact that it steadfastly championed the cause of Boulder Dam even before it became a national issue. Further, The Examiner has been the instrument, which presented the facts in the case, promoted the interests of millions of people and defended popular rights. An influence powerful enough to crystallize a project can be profitably employed by manufacturers and retailers to exploit *their* commodities. Advertising in the Los Angeles Examiner has a background of reader confidence that makes your message doubly effective.

As this advertisement testifies, the *Examiner* was not merely reporting the who, what, when, and where of the project, but instead actively and openly championed the dam and fancied itself "an influence powerful enough to crystallize a project." Although the *Examiner* presents its cause as being the defense of "popular rights . . . in the interests of millions of people," the visual and verbal arguments in this edition clearly serve to maximize and celebrate its own influence over public opinion, particularly with regard to Hoover Dam.

The *Evening Herald* and *Examiner* editions discussed here represent an evolution from grainy, primitive images of Boulder Canyon to richly symbolic illustrations that visualize Hoover Dam as part of an idealized future for the entire Southwest. The illustrations placed within celebratory and even hyperbolic text are mechanisms through which the dam project is being positioned as the lynchpin for the future of the region. By considering the visual and verbal features of these types of illustrations, we can better understand both the ways in which the outlook of the region was presented to the public before the dam was built and how dam imagery intersected broader cultural and social discourses regarding gender, nature, technology, and economic prosperity. These two cases also provide insights into the important role newspapers played in promoting the dam project and are indicative of a pre-construction period in which newspapers were often self-promotional while at the same time offering glowing, imaginative, and magniloquent projections without a hint of negative association or critical commentary.

Ideologies of Knowledge Making

On June 3, 1922, when Representative John W. Summers of Washington stood to address Congress, he used to his advantage an important and ever-present tool in the discourse envisioning the future of the southwestern United States—a map. In examining some of these maps, we can consider how they work as part of a larger visual-verbal discourse of value systems and cultural knowledge making with regard to Hoover Dam. Representative Summers's comical "Jack Rabbits and Markets" speech encouraged extending the powers of the War Finance Corporation Act of 1918 to provide funding for agricultural interests in the West and to promote desert and water reclamation projects:

> MR. SUMMERS: A short time ago on a sagebrush plain in the far West there occurred a jack rabbit drive. Between the hours of 9 and 11:30 A.M. 8,000 jack rabbits were killed. Shortly thereafter another drive in a nearby territory slaughtered 5,000. A third drive resulted in 6,000 fatalities: and still the jack rabbit remains undaunted and millions of unborn jack rabbits are weeping to be born. [Laughter.]
>
> MR. KINCHELOE: Will the gentleman yield?
>
> MR. SUMMERS: A little later.
>
> MR. KINCHELOE: I would like to know why they killed those jack rabbits. I thought they were good to eat?
>
> MR. SUMMERS: They are an expensive pest. . . . Great rivers flow through these useless lands on to the sea, but the fragrant sagebrush plains and the jack rabbit are the same yesterday, today, and—shall I say forever? . . . Will it mean more to this Nation to develop these lands and homes and markets, or shall this decade and this century leave them untouched while tens of thousands of service men beg for farms? . . . Shall the jack rabbit and his progeny forever hold the sagebrush plains? [Applause.][29]

The point of his speech was that, starting with the original colonies, vast sums of money had been invested in the eastern United States. Lands that were once thought to be "useless," from the Appalachian Mountains to the Mississippi Valley to the midwestern plains, had all developed into what Mr. Summers described as a "great producing and consuming empire—the backbone of the Nation." Lands that were formerly the equivalent of the sagebrush plains of the West had begun producing and consuming grain, flour, coal, gasoline, steel, lumber, brick, and cement, all buttressed by a vast system of industry, commerce, and agriculture.

In support of his argument, Representative Summers produced a map on which he showed the points of origin for the "millions of dollars worth of merchandise . . . [shipped to] one of the 26 reclaimed plains" (figure 3.08). Both the map and the speech were favorably received:

> MR. GARDER [of Texas]: I take it that the gentleman's argument is in an effort to show the House that it is wise public policy to continue to reclaim these lands?
>
> MR. SUMMERS: That is my object.
>
> MR. GARDER : I think I can assure the gentleman that there will be a large vote in favor of this bill. . . . I think I can assure the gentleman that there will be three-fourths of the Members on this side of the House who will vote for this bill. According to the map which the gentleman has exhibited here there is only one point in Texas which ships anything into that Idaho project. . . . Not withstanding that fact, I think that a very large proportion of the Texas delegation will vote for that proposition.

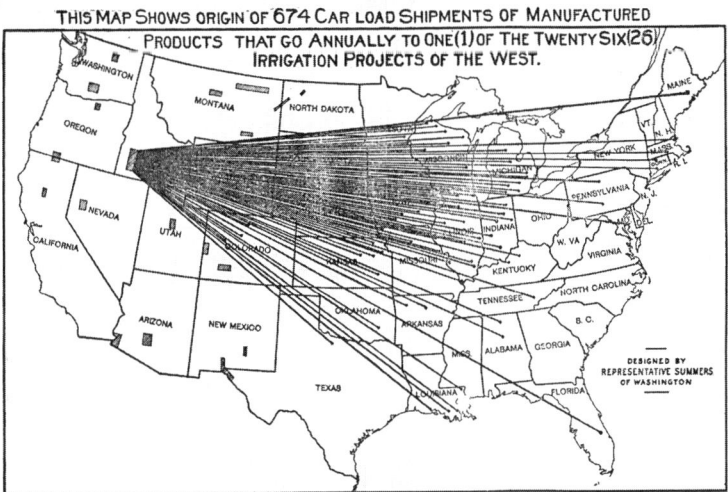

Figure 3.08. "Origin of 674 Car Load Shipments of Manufactured Products That Go Annually to One (1) of the Twenty-Six (26) Irrigation Projects of the West," June 3, 1922 (poster, Congressional Record, US Government Printing Office).

MR. SUMMERS: We of the West have seen the benefits of reclamation to the whole country and are extremely anxious to see this policy extended. [Applause.][30]

The map that Representative Summers used in his speech was a compelling visualization of how economic development in the West was connected, and economically beneficial, to many regions of the country. Mr. Gardner, seeing (literally *viewing* on the map) that the Idaho reclamation project was supplied by just one enterprise in his state of Texas (according to the map), was nonetheless sufficiently persuaded to suggest that a "very large portion" of the Texas delegation would support the legislation. The thick black lines indicating material contributions of many states to just one reclamation project show the vast network of manufactured goods funneled into a program that, although likely far from the minds of lawmakers in eastern states, had a direct connection to their constituents' businesses.

In the same *Los Angeles Examiner* edition discussed earlier, we find a full-page illustrated map of the Southwest labeled "Graphic Forecast of Boulder Dam Possibilities" (figure 3.09), which seeks to depict the potential economic impact to Los Angeles by directing the action of the image (and the viewer's attention) toward Southern California. The trains, airplanes, and ships on the map are speeding toward Los Angeles, indicating that transportation, industrial materials, and natural resources, along with water, will be flowing into, not out of, the area. Next to a train heading south from Nevada, one label reads "Unlimited Supply of Iron Ore From Utah," and another "Power Will Wrest Millions From Vast Stores of Non-Metallic Minerals." The lone image of movement out of California is a train heading away from Imperial Valley labeled "Irrigable Area Capable of Producing $175,000,000 in Annual Crops" and "Early Season Products for the East." Cargo leaving the area represents the vast sums of money generated for Southern California from the sale of agricultural produce—crops that are, of course, dependent on irrigation water from the Colorado River. In the bottom left of the map we see a cornucopia spilling forth bags of money and piles of coins. The cornucopia, labeled "Prosperity," is squeezed between a large gear, representing industry, and a Romanesque building labeled "Bank." Economic fortunes are clearly the primary theme here, as the caption reads "Project Will Develop New Wealth of $12,000,000,000."

Electricity is also a prominent feature in this image. Large towers with power lines connect the dam site to Los Angeles, while a caption above the city reads "Cheap Power Will Make Los Angeles Hive of Industry." To

Figure 3.09. "Graphic Forecast of Boulder Dam Possibilities," *Los Angeles Examiner,* July 16, 1929. Courtesy of the Boulder City Museum and Historical Association.

the north of the city near Owens Lake a small industrial complex caption announces, "Electricity will Release Immense Values in Commercial Chemicals." In a scene to the southwest of the city, rendered in comic book style, a magnet (symbolizing the Los Angeles area) emits electrical bolts that draw buildings—hurtling through the air—toward it with irresistible force. The scene is accompanied by a caption that reads "Power Magnet for Industries." Smaller individual images across the top also suggest that electricity from the dam will bring economic possibilities. On the left, a large turbine and power plant are highlighted by text that reads "1,000,000 Horsepower," while other captions tout additional benefits: "Power Equal to 23,000,000 barrels [of oil] a year," "Electric Railroads," and "Pleasure Lake Formed by Dam" (complete with luxury ocean liner).

The only image that alludes to benefits to be enjoyed by something other than industry, business, or agriculture, and the only circular image on the page, is in the bottom left. The scene shows a woman standing at a kitchen stove. The caption reads "Cheap Electricity in the Home." In this image the woman is deemphasized, dwarfed against a depiction of the entire region and placed in a domestic setting cordoned off from the rest of the scene by the circular frame of the panel.

In addition to the visual features, we should also consider verbal elements of maps that define the values the visuals have established. Text in the *Examiner* map such as "Power Will Wrest Millions from Vast Stores of Non-Metallic Minerals" promises untold wealth for Southern Californians and is a powerfully reinforcing verbalization of the optimistic worldview expressed in the map. This perspective is buttressed through verbal labeling that extends the economic imperatives of both the news media and Southern California's agricultural and manufacturing industries. In this map we see effusive labeling suggesting the boundless economic gains that will come from the Boulder Dam project: "Unlimited supply of Iron Ore from Utah," "Electricity Will Release Immense Values in Commercial Chemicals," "Cheap Power Will Make Los Angeles a Hive of Industry," "Water for 7,500,000 People," and "Cheap Power Will Develop Great Arizona Mineral Resources."

We have now seen several maps used to represent and promote the Boulder Dam project in its initial stages. In the Bureau of Power and Light brochure we saw Hoover Dam towering over the entire Southwest and delivering water to Nevada and Southern California. We have also seen a full-page map of the Southwest in the *Los Angeles Examiner* that highlights the benefits of the dam project for Southern California's economy and the city and citizens of Los Angeles. In Representative Summers's map of the United States we see the connection between manufactured goods from across the country and a reclamation project in Washington State. These maps, as a genre of visual and verbal representation and persuasion in the discourse about the dam, call us to consider mapping as part of an ongoing motif of social knowledge making about the project.

The term *map* is often used to describe the relationship between a visual representation of geographical features and that which is represented: a diagrammatical representation showing the relation of one location to another. Maps are often accepted as convenient, impartial, objective, and accurate reproductions of roads, political boundaries, topography, and so forth. But maps are, at their essence, a part of a cultural system and should be understood as ideological sites of visual-verbal emphasis within a wider discourse of representation.[31] They are also social constructions that are historically relative and need to be understood in the context of their own time period as one of many possible representations of space and place.[32]

Maps are also a particularly good example of visual-verbal representations infused with ideology. Considering maps as such cautions us against accepting a given map as a wholly objective, obvious, commonsense diagrammatic representation of reality. Instead, we should think of maps as ideological tools, as embodiments of cultural and social conventions resulting from a complex arrangement of economic, social, and political positions, and as mechanisms that operate to mask realities of power and solidify social order. In short, visual and verbal information contained within maps should be viewed in relation to its sociohistorical context and changing relations of power and knowledge.[33]

Understanding maps as ideological tools can allow us to do what Barton and Barton describe as moving "beyond the study of visuals as embodiments of cultural conventions to the study of the ideologies that, in turn, inevitably underlie those conventions."[34] Centering on what they term *rules of inclusion* and *rules of exclusion,* Barton and Barton argue that maps are mechanisms by which groups (nations and other geopolitical entities) express possession of, and thus legitimate, names and places.[35] For them, maps serve the dominant interests by focusing on inclusionary and exclusionary ways of knowing through representation in which, for example, "we never see slums, buildings in poor condition, or suggestions of danger." Maps, they argue, often present "an optimistic worldview, an image which focuses on only the positive aspects of the urban, and in some instances, rural life."[36] J. B. Harley, one of the formative figures in the study of the history of cartography, argues that maps show not only the physical but also the human landscape in which the rules of social order implant themselves as codes:

> Pick a printed manuscript map from the drawer at random and what stands out is the unfailing way its text is as much a commentary on the social structure of a particular nation or place as it is on its topography. The map maker is often as busy recording the contours of feudalism, the shape of a religious hierarchy, or the steps in the tiers of social class, as the topography of the physical and human landscape. . . . The rule seems to be "the more powerful, the more prominent." . . . Using all the tricks of the cartographic trade—size of symbol, thickness of line, height of lettering, hatching and shading, the addition of color—we can trace this reinforcing tendency in innumerable European maps. We can begin to see how maps, like art, become a mechanism for defining social relationships, sustaining rules, and strengthening social values.[37]

But, other than looking at names and labels on a given map, how can we go about decoding who or what the dominant interests are? Naming performs the role of legitimating dominant social interests, but so too do learned strategies of representation for particular phenomena, such as one based on hierarchy of space. Space on a page, for example, is not isotropic. Instead, we perceive it as anisotropic, or having unequal value depending on where elements are placed on the page. Elements placed anisotropically become part of a strategy of privileging by positioning.[38] One well-known anisotropic arrangement is through *centering*. Objects placed at the center of an image privilege that object—the man at the center of the *Examiner* illustration, for example, or North America or Europe at the center of the Mercator map.

Another anisotropic placement privileges the *top*. We often see this in organizational charts, for example, where the CEO is on the top and the power of each position decreases as the boxes descend down the page, or in newspapers where the most important news story of the day is headlined at the top of the front page. In these examples, the top has been reified to a point that we naturally assume that it is the most important. In the *Examiner* map, we see the headline at the top: "Graphic Forecast of Boulder Dam

Possibilities." In saying it represents a "forecast," the *Examiner* is suggesting that this is an empirically informed prediction of what is likely to happen— their advanced indication or calculation of future events. Directly below the headline are the series of boxed illustrations I described previously: a one-million-horsepower turbine producing electricity; luxury cruise liners and tall skyscrapers; electricity the equivalent of twenty-three million barrels of oil; electric trains; luxurious pleasure boating on the dam-created lake; and fifteen thousand jobs, all resulting from the dam's construction. Their placement across the top of the page directly below the headline suggests an important link between economic and electric power and the Boulder Dam project.

These visual-verbal arguments suggest increased prosperity with no downside, but prosperity for whom? We have already established that the *Los Angeles Examiner* itself was unabashedly championing the project, and even though it would seem to be a powerful interested party, its intentions are hardly hidden. But how might the *Examiner's* attempt to use the map as a persuasive tool be linked to claims of power and authority? Hydroelectric power generation is seemingly egalitarian, as it would provide cheaper electricity for the consumer, as well as profits for the utility. Mass transportation in the form of electric railcars would seem to benefit the lower and middle classes. Pleasure craft on the lake, as depicted, would seem to benefit the wealthy, while construction jobs would benefit the laboring class. Each of these small illustrations contains an in-box caption describing the benefit depicted. One of them, however, does not. One box shows a luxury ocean liner backgrounded by a skyline of tall buildings, similar to those in the *Evening Herald* image. These symbols of modernity and affluence have no relationship to Hoover Dam itself, but instead insinuate that the dam project is part of a larger move toward regional power and prosperity. These symbols of prosperity—the machinery of industry, commerce, and leisure—in this image are vital because the overall argument rests on regional prosperity and not on the dam itself. Indeed, the dam is barely depicted; its placement is highlighted by the large bolded text "Boulder Dam" as opposed to any pronounced visual cues. These symbols are what Michel de Certeau calls "narrative features."

In *The Practice of Everyday Life*, de Certeau writes, "The machine is the *primum mobile*, the solitary god from which all the action proceeds."[39] Imagery of industry, machinery, and transportation fills the *Examiner* map.[40] At the top of the page we see illustrations of ocean liners, turbines, oil derricks, pleasure ships, and electric trains, all products of the dam serving to benefit the industry and citizenry of Southern California. For de Certeau, "narrative features" of a map such as ships, buildings, animals, and other characters are not inconsequential. Instead, these figurations indicate military, commercial, and political operations and provide the story—the cultural history—that underscores the cartographic functionality of a map: "Far from being 'illustrations,' iconic glosses on the text, these figurations, like fragments of stories, mark on the map the historical operations from which it resulted. . . . But the map gradually wins out over these figures; it colonizes space; it eliminates little by little the pictorial figurations of the practice that produce it. . . . The map thus collates on the same plane heterogeneous places, some *received* from a tradition and others *produced* by observation."[41]

Although de Certeau laments the gradual loss of certain narrative figures, they nonetheless remain a forceful component in maps of Hoover Dam. Railroads are a prominent example of narrative figures in the *Examiner* map and were often included as visual accoutrements in many early maps of the dam project. A 1927 booklet, *The Colorado River Project,* issued by the Boulder Dam Association of Los Angeles, includes a two-page map spread showing the irrigable areas below Parker Dam. However, the only features included in the map other than political boundaries of the states are the Colorado River and its tributaries, irrigable lands in Southern California, and railroad lines that crisscross the Southwest. Early maps from the US Bureau of Reclamation showing the All American Canal route and panoramic perspectives of the dam project also typically include major railroad routes throughout the Southwest. In the *Examiner,* the three trains that crisscross the map divide its graphical features and parallel the electric transmission lines, helping to direct the viewer's eye toward Los Angeles. The trains show the origin, means of transportation, and destination of the raw materials that will provide wealth to California. Regarding the railroad, de Certeau comments, "There is something at once incarcerational and navigational about railroad travel; like Jules Verne's ships and submarines, it combines dreams with technology." He continues that machinery such as trains "not only divides spectators and beings, but also connects them."[42] The *Examiner's* trains divide the map, but also bring energy and movement, symbolizing the promise of wealth and prosperity through technology. They provide what de Certeau describes as the "fragments of stories" and place a historical "mark" on the map.

The railroad, like the dam, was a symbol of technological progress and modernism: the "iron horse" riding across the plains delivering people and goods farther, faster, more cheaply, and in more quantities than ever before. The railroad also symbolized Americans' relationship with the landscape and often appears as a metaphor to describe that relationship. In *Walden* (1854), Thoreau wrote of the binary nature of the railroad, in which, on one hand, it did have some redeeming qualities, but like other technological advances it made life both easier and more complex. The "fiery beast," as he called it, brought material benefits, but also represented the end of agriculture and the destruction of nature. In *Machine in the Garden*, Leo Marx describes how the railroad had become an American national obsession by the mid-nineteenth century, and he uses the railroad as his primary metaphor to describe the unalterable fate of technology destroying the pastoral. "It is the embodiment of the age," he writes, "an instrument of power, speed, noise, fire, iron, smoke—at once a testament to the will of man rising over natural objects, and, yet, confined by its iron rails to a predetermined path, it suggests a new sort of fate."[43]

The railroad also serves as one of several narrative figures in other maps of the dam. In two maps published in the early 1930s we see an evolving scene in which nature, technology, and culture coexist in a fanciful imagining of the West. In the 1931 booklet *White Gold: The Story of Hoover Dam*, issued by the American Steel & Wire Company, we see an illustration of the Southwest as a menagerie in which animals, nature, urban zones (Salt Lake City, Phoenix, and San Diego), and technology (depicted by two trains speeding across the landscape) populate the region (figure 3.10).

Here active nature is depicted as a disembodied winged cherubic head blowing wind across the region. Nature is also represented as mountains and desert landscapes, clouds, cacti, and the Colorado River. This image is also one of the very few related to Hoover Dam in which we encounter animals. Whereas the few animals that are shown in dam imagery elsewhere are either pack mules, horses ridden by men, or fish being caught out of the lake, these animals—bear, deer or elk, coyotes, a horse—are depicted as free of human control and part of the landscape. There is, however, a visual-verbal disconnect as the caption, "In the Center of Millions of Acres

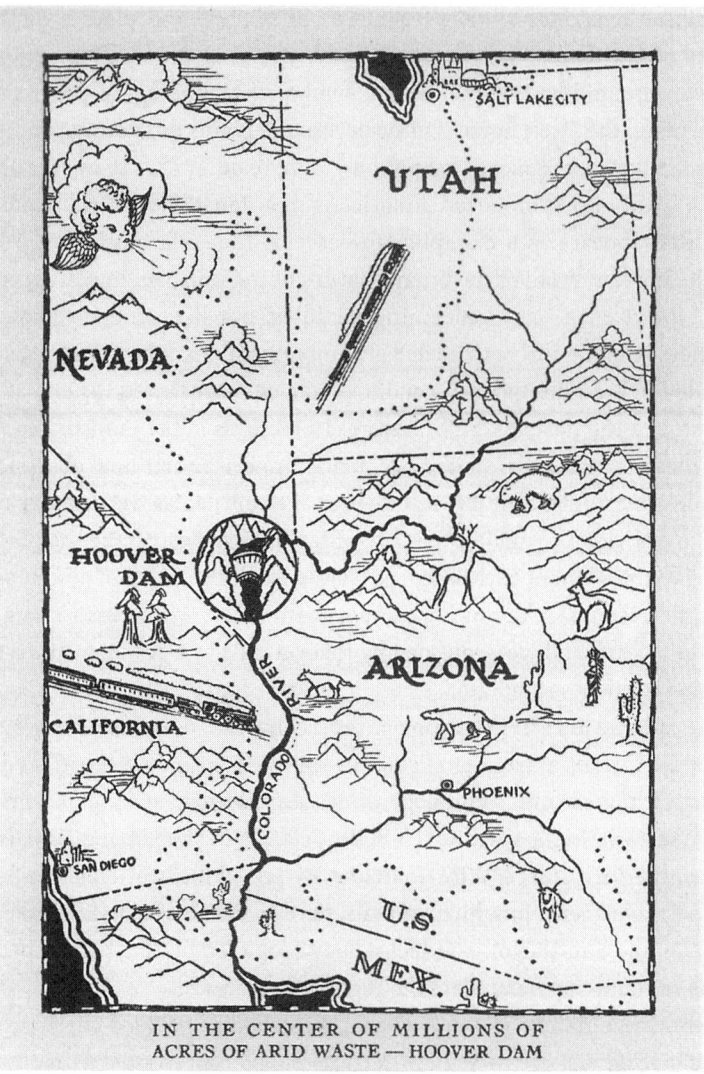

IN THE CENTER OF MILLIONS OF
ACRES OF ARID WASTE — HOOVER DAM

Figure 3.10. "In the Center of Millions of Acres of Arid Waste—Hoover Dam," 1931 (in *White Gold: The Story of Hoover Dam,* American Steel & Wire Company of New Jersey). Courtesy of the Boulder City Museum and Historical Association.

of Arid Waste—Hoover Dam," ruptures this initial reading. No longer a scene of coexistence, the text redirects our reading of the image from tranquil menagerie to barren wasteland. The lone human figure depicted in the image intimates this reading as the crudely drawn figure is dressed in what seems to be pioneer clothing or native dress. Either way, in its unusual context the figure suggests an unruly, wild, unexplored landscape.

A nearly identical iteration of the same map appeared in *Popular Mechanics* the following year, yet with a slightly refined depiction. No longer a crude rendering, the figures are more elegantly illustrated, and the image is multilayered, with both dark and light lines tracing the Western landscape. Again we see animals—a wolf, an elk, a mountain lion, a mule, and a fish—shown as part of the menagerie of the West along with the disembodied cherub-faced wind, the mountains, railroads, and cityscapes. But this time, the image features more symbols of advanced transportation and cosmopolitan culture: cars and planes accompany the railroads, while beachgoers and filmmakers appear near Los Angeles. Instead of a pioneer or Native American, a prospector and his mule reinforce the "White Gold" theme of the article.

When we look to what is included or emphasized in a map—the strategies and devices used to map that can point to either overt or covert claims to power—we can determine not just what is mapped, but what is represented.[44] Here, the *White Gold* and *Popular Mechanics* maps do not succumb to the complete cartographic colonization de Certeau laments. Instead, these maps use the narrative figures in the service of the map's primary function: visualizing an imagined future of the Southwest. These figures of landscape, industry, transportation, and culture are a result of historical and political operations, but they also tell the tale of what could be, or would eventually be, reality.

Another *Popular Mechanics* article in the March 1931 issue titled "A Glimpse into the Future at Boulder Dam" claims that such imagining involved not just what the dam might look like, but how the dam would alter the future reality of the region. Whereas the figurations de Certeau describes worked to chronicle the military, architectural, and political aspects of a culture, maps of the Boulder Dam project ultimately function as both chronicler and producer of notions about the dam project and its role in an imagined vision (and visualization) of the future.

As we have seen, electricity, transportation, and agriculture are the dominant themes of the visual narrative figurations illustrated in the *Examiner* map. But an additional feature of the map is not quite as explicit, and is one that I have alluded to several times already. There is a palpable yet only implied sense of regional and egalitarian prosperity expressed in its visual and verbal signifiers. Within the particular context of the Boulder Dam project we see that the pursuit of prosperity for both industry and individuals serves as a powerful position from which to form persuasive arguments in favor of the dam's construction. A feature map, by focusing on what is desirable, furthers a boosterist agenda for the sole purpose of creating a positive image, what Barton and Barton call a "self-indulgent morality of capitalist consumerism."[45] The imagining of desirable outcomes for the Boulder Dam project in the *Examiner* map and other promissory pre-construction images are cases in which the ideologies of Americanism—capitalism, the American dream, and so forth—coalesce in a synchronic relationship in the service of the Hoover Dam project and its dominant ideologies.

ALTHOUGH ITS EARLIEST REPRESENTATIONS relied on verbal descriptions accompanied by depictions of simple contours, shaded drawings, and superimposed data sets to communicate its size, scope, location, and appearance, the imagery of Hoover Dam became increasingly sophisticated as regionalized, stylized representations of cultural ideologies took on more complex and subtle forms. From the militaristic depictions that suggest a campaign of man and technology against the menacing Colorado River to subtler yet no less ideologically powerful maps, illustrations played an important role in establishing the dam as a central feature in the grand narrative of a future Southwest reaping the benefits of cheap electricity and abundant irrigation and drinking water.

The illustration itself is enlivened by virtue of its being unencumbered from any temporal restriction of depiction, something that gives it a distinct advantage over photography: illustration frees artists to conceive of a becoming-real, a future event, a grand narrative full of metaphor, allegory, and allusions to what the dam would mean for the Southwest. Whereas illustration (and other renderings by human hand such as paintings) can imagine an alternate reality, depict something that is not in the direct experience of the artist, and project the future, photography (excluding the aid of postproduction image manipulation technologies) can only capture what is indexical, and can only picture the past. But given these freedoms of depiction that illustration is provided, the sheer size and scope of the project still frustrated artists in their attempts to depict the dam, thus requiring them to rely on grand verbal explanations to support their images. Once actual construction began, however, and photographic documentation became the primary means of chronicling the dam's progress, we see a distinct turn toward efforts to capture in images what had been previously articulated primarily in words. It is this effort to depict the sublimity of the dam to which we now turn.

4

Picturing the Sublime

In a 1937 promotional brochure, the Chamber of Commerce of Boulder City, Nevada, described the "bold engineering genius of [Hoover Dam's] design" as a "world wonder" that "has thrown this impregnable fortress against the forces of a devastating river which for centuries had wastefully spent its power." More importantly, it concedes that the ways in which the dam is described—often verbally through the use of statistics, comparisons, and adjectives—is simply an inadequate means of imparting a full understanding of the enormity of the undertaking. In its travel promotion pitch, the brochure argues that words "fail to utterly convey a conception of the magnitude . . . the sheer grandeur . . . of this mighty barrier flung against the torrential floods of the Colorado River. Like many another of the world's wonders, Boulder Dam must be realized" (ellipses in original).

This pronouncement about the dam is far from unusual. Before, during, and after its construction, thousands of similar assessments have been offered in newspapers, pamphlets, booklets, government documents, magazines, cartoons, advertisements, political speeches, radio programs, films, television shows, and more. The description in the Chamber of Commerce brochure encapsulates many of the major themes of those assessments: the economic impact of the project, its symbolism of American engineering prowess, its domination of a treacherous and improvident river, the wastefulness of unaltered nature, and the dam's undeniable sublimity. Moreover, like the Chamber of Commerce, those who have attempted verbal descriptions or visual depictions of the project ultimately suggest that words or images simply cannot do justice to the real-world, in-person experience of visiting the dam.

An article in the October 3, 1931, edition of the *Las Vegas Review-Journal* is typical of the attempts to express the inexpressibleness of the dam. "The greatest change that the brain of man has ever devised in the map of the world," it says, "is being prepared in Nevada today. . . . Facts submitted about the Hoover Dam enterprise to the Appropriations committee of the House [stated] that nearly every dimension is superlative." The article concedes that the dam is more than mere statistics. "Pick what you will about the facts and figures," it rhapsodizes, "[the dam project] is the biggest job in the world today; and the strength of this statement *possesses you*" (my emphasis).

The facts and figures are, nonetheless, impressive. The first batch of concrete was poured on June 6, 1933, and continued unabated twenty-four hours a day, 220 cubic-yards an hour for nearly two years until the last batch was poured on May 29, 1935. All day and night, 365 days a year, railcars delivered a continuous supply of concrete from nearby, custom-built "lowmix" and "himix" concrete plants to one of five cableway systems strung across the canyon. The cableways that hoisted men, machinery, and materials from the rim to the riverbed nine hundred feet below and back lowered the huge, eight-cubic-yard buckets of concrete scooped from the railcars down into the 50 x 60 foot forms where men nicknamed "puddlers" tripped the safety latches to release the slurry. The puddlers then evened out the concrete into the forms using shovels, bars, and vibrating hydraulic prods to evacuate all of the air bubbles. After delivering their payload, cableway operators hoisted the buckets skyward at three hundred feet per minute, only to repeat the cycle countless times. In the end, four and a half million cubic yards of concrete was poured in just under two years, completing in an astonishingly short time what is, to many, the greatest and perhaps most significant engineering project ever undertaken in the United States.

Since construction continued around the clock, some of its most dramatic depictions were of the dam at night (see figure 4.01). Hugh Blair in his 1827 *Lectures on Rhetoric and Belles Letters* wrote that night scenes are generally the most sublime, and commentators witnessing the otherworldly scene of the nocturnal construction site would likely concur. *Fortune* magazine reported that in order for the work to continue at night, the canyon was "lighted like a theatre with incredible clusters of sun arcs bought from a bankrupt San Francisco ball park." Don Black describes his experience of

Figure 4.01. "Night photograph of power house construction," June 19, 1934 (photograph, Ben Glaha, Bureau of Reclamation). Courtesy of the University of Nevada, Las Vegas Libraries, Special Collections. W. A. Bechtel Collection.

viewing the nighttime construction scene in the June 1932 issue of *Popular Mechanics* this way:

> They call it the most spectacular sight in America today. Blackest night above, inferno below, and you perched halfway between on the brink of the abyss, reaching for the stars to keep from falling. It is the fiery pit of haunted dreams, aglow with light and colorful reflection that bounces back and forth between the narrow walls like the echo. Dazzling searchlight, groping restless; headlamps streaming out of nowhere to race on hidden trails and vanish like a dream. . . . And this is the beginning of Hoover Dam—a test of man, machinery and engineering without precedent in all the history of the world.

Joseph Stevens, historian and author of one of the best-known works on Hoover Dam, describes the sight of the dam at night in similarly breathless terms. He writes that the nighttime scene is "unearthly," and when the giant floodlights come to life, they illuminate the concrete "in a blaze of dazzling golden light." He continues, "Confronting this spectacle in the midst of emptiness and desolation first provokes fear, then wonderment, then finally a sense of awe and pride in man's skill in bending the forces of nature to his purpose."[1]

Viewing such a sight in the 1930s—when the world had seen nothing like it before—required the suspension of disbelief, a fact not lost on either government or private producers of Hoover Dam publicity materials. The dam was, to most, a sublime technological triumph that quickly became the subject of media and popular veneration. Headlines such as "Men Move Mountains, Make Way for Greatest U.S. Project—Boulder Dam," which ran in the June 9, 1931, *San Francisco Examiner,* were typical, as other stories proclaiming the dam project to be the "biggest," "greatest," most ambitious engineering project in history ran in newspapers across the country throughout the dam's construction. The Department of the Interior produced a radio program titled "Man Is a Giant" that sought to "vicariously gratify" the "forbidden thrill of visiting the structure" after tours of the dam were canceled during World War II.[2] Hearst Metronome and International Newsreels showed before many of their theatrical releases footage of the dam's construction with overdubbed inspirational music and narration.[3] The revered University of Notre Dame football team, after their tilt on the gridiron against the University of Southern California in December 1936, stopped off at Hoover Dam to pose for press photos in their uniforms and three-point stances, seemingly ready to rush the camera while the dam towered behind them in the background. The Mormon Tabernacle Choir even bestowed their liturgical honorifics to the project's completion through a live radio broadcast performance from atop the dam on July 29, 1935.[4]

Government agencies, not to be outdone in attempts to glorify the dam, used their own inspirational language to describe for the public the enormity of the undertaking. A 1930s Bureau of Reclamation article (originally printed in *Compressed Air Magazine*) titled "Huge Trailer Hauls Penstock Pipe at Boulder Dam" offers a typical example. It states, "Because Boulder Dam is the largest structure of its kind ever built, it logically follows that its construction calls for doing various essential operations on an unprecedented scale and with equipment of unusual size." The article goes on to describe aspects of the construction project as "huge," "record size," "faster and in greater quantities than has ever been done before," and "surpass[ing] in magnitude all others of their respective kind that have ever gone before."[5]

But before considering in more depth some trends of sublime representation with regard to Hoover Dam, I should first discuss why the sublime is an important notion in the image-text relationship, and how there seems to be something of a paradox of the sublime: rather than leaving us speechless, as most theories of the sublime have claimed, I argue that the sublime instead does the opposite—it compels humans to attempt to represent their experiences of it in both prose and images. In particular, I want to consider how images have been employed to convey sublimity with regard to Hoover Dam.

Paradox of the Sublime

The recent emergence of visual studies has been, in large part, a response to two states of affairs: first, the structuralist enterprise led by Claude Lévi-Strauss that sought to use "signs" to describe culture as "texts" that can be "read"; and second, the traditional dominance of the verbal over the visual as the apotheosis of human intellectual achievement. Although structuralism's precepts were exposed as being contradictory to what the endeavor actually claimed, and the study of the visual has moved toward more nuanced post-structuralist and postmodernist views, some scholars today still rely heavily on structuralism's basic premises.[6] Several of these scholars have set about constructing a highly sophisticated hermeneutic of "reading" images in which following certain rules and codes would allow us to turn our traditional understanding of the image-textual relationship on its head: instead of images supporting or illustrating the meanings and arguments in the text, the text now supports and augments the meanings of the image. As a result, they propose that we should learn not just a grammar of text and language but also a "grammar" of images so that citizens in our new visual era can "read" an image in the way that we have traditionally read text.[7]

This tendency for some to adhere to the linguistically based structuralist conception of "reading" images and analyzing signs as a way to determine meaning in society, however, overlooks the very two things that make images distinct from text: first, the capacity of images to immediately convey an enormous amount of information that would require pages of text to approximate yet would ultimately fail to replicate; and second, the capacity of some images to have an instant and visceral impact on a viewer that verbal text simply cannot duplicate.[8] This visceral impact can, in some cases, be taken to such heights as to exact an undeniably primeval response that moves us to a state of intense, awe-inspiring emotion, a state often referred to as the sublime—the very state that has left writers and artists covering Hoover Dam's construction struggling to describe the indescribable and to visualize the unvisualizable. However, the concept of the sublime also speaks to a fundamental paradox in much of the dam project's documentation, namely, that writers and artists were tasked with expressing to the public their breathtaking encounter with the dam site, an experience that was also routinely described as being indescribable.

The front page of the October 1, 1935, *Los Angeles Times* neatly encapsulates this paradox in its coverage of the dam's dedication address delivered by President Franklin D. Roosevelt with the headline "Dam Awes President," while its subheading reads, "'I'm speechless,' Roosevelt says as He Gazes at Gigantic Structure." The poignant irony here is, of course, that although the visual spectacle of the dam is beyond words and rendered the president of the United States "speechless," its very sublimity called for a speech to honor the dam and those who built it. Indeed, the delivery of a speech was the only reason the president was there.

Many others have attempted to describe the indescribable characteristics of the dam. Richard Guy Wilson wrote that "Hoover Dam, like all great works of art and all important symbols, transcends description and interpretation. It is more than what can be written or expressed."[9] Similarly, in explaining how visitors viewed the construction progress from the rim of the canyon, Marc Reisner wrote, "There was usually a long moment of silence, a moment when visitors groped for something to say, something that expressed proper awe and reverence for the half-formed monstrosity they saw. The dam defied description, it defied belief."[10]

Yet, description and expression were exactly what writers and artists were tasked—in fact compelled—to produce throughout the project. Whether conveying through images what millions of cubic yards of concrete looks like, or putting into words what the stifling heat of the desert feels like, chroniclers of the dam project (whether they were officially sanctioned writers, photographers, or artists, or simply visitors to the site) were faced with portraying a sublimity that, as these comments testify, resists representation and immobilizes humans in its presence, thus inhibiting the functional capacity of communication itself.

Much like the Chamber of Commerce assessment in which the dam must be "realized" to appreciate its true awesomeness, Enlightenment and romantic reflections on the sublime, as well as many recent scholarly theories of the sublime, the American sublime, or the technological sublime, emphasize the importance of direct and unmediated experience or perception of the event, object, or landscape. Gregory Clark, for example, writes that American sites such as New York City, Yellowstone National Park, and the Grand Canyon work rhetorically by leading individuals to identify with particular symbolic places at which visitors gain from the rhetorical experience of landscape a sense of national identity.[11] What Clark describes is an example of how individuals form a sense of community identity through the lived experience of visiting or moving through particular American landscapes.

I would argue that Hoover Dam, too, is a symbolic place whose rhetorical experience generates a sense of national identity. As such it not only strikes us "dumb with amazement" but also has what David Nye describes as a "transcendent importance both in the lives of individuals and the construction of a culture."[12] However, in a diverse American population stretched across millions of square miles with yawning socioeconomic and cultural differences, such sites would seem to also resist any communally lived experience. For the majority of people, encounters with these landscapes are not in situ, but are instead mediated primarily through various verbal and visual representations. Consequently, we must consider not only whether it is possible to experience the sublime through representation, but also the effects on something's sublimity when subjected to widespread and even ubiquitous mediated representation. As I have mentioned, thousands of images of Hoover Dam were circulated to the public in a variety of media formats over a long period of time. Furthermore, I have argued that this supradiscursive imaging process was integral to the development of the cultural mythology, indeed the very iconicity of the dam. Thus, the images of the dam and their verbal accompaniments reveal the ways in which chroniclers of the project sought to produce a testimony of their own bodily witnessing of the sublime. The value in these types of visual-verbal representations, then, is in the chroniclers' communicating the rhetorical act of witnessing, which attempts to replicate their experience of sublimity through acts of representation. What I argue on the pages that follow, then, is that rather than "realizing" the sublime, the act of attempting to articulate the experience through visual-verbal representations instead opens the possibility of the viewer/reader achieving something akin to its state by compelling a reassessment of the *relationship* between themselves and something greater than themselves.[13]

The Sublime in Western Thought

The sublime has a long history in Western thought and has received something of a revival in contemporary philosophy, art history, and literary studies, as conceptions have stretched to include not just traditional notions but also apocalyptic, democratic, technological, economic, urban, and maternal sublimes, to name just a few. The first theoretical discussion of the sublime appears in *Peri Hupsos* (sometimes spelled *"Peri Hypsous"*), a treatise historically attributed to one Cassius Longinus but which is now considered to be from an unknown author of the first or third century C.E., although most still reference Longinus or "Pseudo Longinus" as the author for the sake of simplicity and to avoid confusion. The Greek term *hupsos* has a literal meaning of "elevation" or "grandeur," but has traditionally been translated as "the sublime" or "sublimity."[14] For Longinus the sublime is not a static product inherent in discourse; rather, it is an affective response to the finest oratory and writing inspiring awe and veneration in its auditors. Although clearly informed by the rhetorical theory of his day, Longinus is not constrained by its tenets, nor is he merely interested in techniques of persuasion.[15] In fact, he describes sublimity as a "spell" that prevails over persuasion or gratification. To him, we can control our emotions with regard to persuasion, but "the influences of the sublime bring power and irresistible might to bear, and reign supreme over every hearer."[16]

Other important developments of the theory of the sublime occurred in the eighteenth and nineteenth centuries. In the wake of the appearance of Boileau's translation of Longinus's treatise, the sublime found its way into the mainstream of European aesthetic thought where several major Enlightenment thinkers, most notably Edmund Burke and Immanuel Kant, significantly expanded its precepts. In *A Philosophical Inquiry into the Origin of Our Ideas of the Sublime and Beautiful,* Burke broadens the concept beyond verbal discourse by arguing that sublimity lies not only within literature, art, and oratory, but also in the grandeur of nature. It is present in experiences of darkness, emptiness, vastness, starry nights, thundering waterfalls, raging storms, and roaring animals. Burke also offers that the intense emotional response to the sublime is a reaction not only to the experience of pleasure and beauty but also to pain, ugliness, terror, and danger. In Burke's sublime, the object so entirely fills the mind that it cannot entertain any other thought and cannot reason on the object that fills it; its effect is astonishment to the highest degree. For Burke and other early modern writers, the sublime was regularly associated with the grandeur of nature, the experience of which could not be reduced to verbal discourse. Accounts of the sublime often focused on a viewer's encounter with a physical object or vista in which the sublime induces initial incomprehension that holds the viewer's gaze and imagination while inspiring feelings of amazement, terror, wonder, awe, or marvel because one has no previous experience with which to compare it.

If the natural sublime of the Enlightenment and romantic thinkers regularly led to awe and marvel at God-as-creator, we see a radical shift when the agent responsible for the sublime experience is a human. The awe and veneration previously reserved for God-as-creator is redirected to the man-as-inventor with his (apparently) unlimited capacity to create and order nature as he sees fit.[17] In a striking passage in *The Critique of Judgment*, Kant writes that in the "immeasurableness of nature and the incompetence of our faculty" we find our own limitations, but that we also find in our own minds "a pre-eminence over nature even in its immeasurability."[18] Just as the might of nature forces our recognition of our own finitude, for Kant it also reveals ourselves as separate from, yet superior to, nature. He writes, "Therefore nature is here called sublime merely because it raises the imagination to a presentation of those cases in which the mind can make itself sensible of the appropriate sublimity of the sphere of its own being, even above nature."[19] Here Kant argues that although the sublime causes us to become aware of our own physical feebleness, we are not to "bow down" before nature. Instead, sublime nature causes us to recognize our independence and preeminence *over* nature.

The Ubiquitous Sublime

Following these conceptions, there have been numerous other commentaries on sublimity from Hegel, De Man, Eagleton, Jamison, Lyotard, Marx, Derrida, and many others. Within all of these commentaries several themes emerge that allow us to regard a few fundamental attributes of the sublime. First, sublimity is punctuated by the affective response of a spectator whether through pleasure, awe, pain, terror, triumph, self-awareness, or any other sense of realizing one's finitude in the face of the infinite. Second, it defies, transcends, or stymies discourse; it strikes people "dumb" and is at its essence indescribable, immeasurable, and undenotable. Third, it resists representation because of either the impracticable aptitude that would be needed to reproduce it or the exceptional qualities of the event or object. And fourth, it is necessarily rare—one cannot be stunned to silence by something if one encounters it on a daily basis, and thus infrequency is a requisite for the surprise and awe that one experiences. If nothing else, the sublime seems to be these things.[20]

Yet, given these attributes, when we consider the many visual and verbal representations of Hoover Dam, we can see that the dam, although claimed by writers, politicians, engineers, journalists, artists, and others to be a sublime undertaking, seemingly violates every requisite just described: it did not strike people dumb, did not eliminate discourse, did not resist representation, and was not diminished by viewing the site day after day. In fact, if anything, it inspired discourse and representation on a level unequaled for any engineering project in history. Moreover, if we consider that there are thousands of images and articles expressing the inexpressible nature of the project, we see that attempts to convey its sublimity through visual-verbal representation are not rare, but everywhere.

We might ask, then, whether the sublime is really so different from some other human experience such as love, for example, which is said to render one speechless, is rare, and has affective responses, yet has inspired countless visual and verbal musings and representations. Discourse can certainly be found lacking when one is in rhapsody from a religious experience, in the presence of a paramour, awed by a natural wonder, or overwhelmed by the boundless expanse of the universe, yet we can inventory the thousands upon thousands of visual and verbal attempts at representations of each. While Longinus indentifies the effect of the sublime as beyond definition, he nonetheless immediately proceeds to outline the five rhetorical elements that an orator can use to bring about such an effect. Thus, we might go so far as to say that humans are, in fact, compelled to verbalize and pictorialize the very experiences that are generally considered beyond conveyance.

The sheer volume of Hoover Dam imagery and its description as a sublime engineering achievement suggests that the sublime is not beyond verbal or pictorial representation any more than any other type of human experience, or at the very least has not dissuaded endless attempts to do so.[21] Instead, these representations function as the rhetorical act of witnessing the sublime, or, in other words, of opening a discursive space for the possibility of the impossible, of the inexpressible, or of the unpresentable. Consequently, we might instead conceive of the sublime as something like our psychological capacity for astonishment, whether it be through "raw" or mediated experience.

However, it is important to note that although there was undoubtedly an inclination to communicate the sublime experience of seeing the dam by way of rhetorical acts of witnessing and mediated representation, what we cannot claim is that these representations allow us to access the "thereness" of a person *actually* witnessing or experiencing the sublime, which, conveniently for us, current theories of representation do not claim anyway. For example, photography was once considered to provide access to unmediated copies of "reality" that would provide us authentic knowledge and "truths" about the world. Technologies such as cameras were thought to be a more reliable source of knowledge than humans because of the shared belief that photos rendered by machines are more truthful recorders of events than people. This belief was eventually discredited, as Roland Barthes, for one, considered photographic truth a myth, not because the photographs aren't true, but because to him truth is always culturally determined and has no mechanism for independent verification. In other words, all experiences escape "true" representation; thus, there is always an absence implicated in any form of discourse regardless of whether it pertains to our experience of love, a tree, a swim on the beach, a photograph of our mother, a grain of sand, a waterfall, or a page of prose.

Following trends in other fields such as linguistics and literary and art criticism, contemporary photographic scholarship maintains that photographers do not simply copy nature or record an objective truth. Instead, a photograph is to be conceived as a *cultural* object—a historically constituted product of human labor and decision making. Photographs of Hoover Dam are technologies (negatives soaked in chemicals and rendered in darkrooms, printed by machines, distributed in newspapers and brochures, and so forth) mediated through other technologies (cameras) and selected by human agency (people decide where to aim the camera, what to exclude from the photograph, when to release the shutter, and so forth). Photographers interpret their surroundings, making choices about which images to record and which to ignore and processing the visual field in front of them before organizing their impression in a physical act of seeing and depressing the button on the documenting machine (the camera).

This is all to say that although the visual-verbal representations of the dam cannot deliver the actual bodily experience, the "thereness" of standing in front of or "realizing" the dam (or the sublime), writers and artists nevertheless endeavored in various inventive ways to do something close to it, or, alternatively, to do something different but similarly impressive—that is, impressing someone's lived experience onto a viewer's mind by showing or describing the *relationship* of a human to the sublime, what Lyotard describes as perceiving our own limitations within a "pictorial or otherwise expressive witness to the inexpressible."[22] The question that remains, then, is *how* did chroniclers of Hoover Dam do this without the viewer actually experiencing sublimity? The rest of this chapter will be an attempt to answer that question.

The Gigantic and the Miniature

While the awe that Hoover Dam inspires has never failed to prompt visitors to try to put their reactions into words, an equally strong impulse was to attempt to capture the experience in images—particularly photographs. In fact, efforts to capture the astonishing visual appearance of the dam, and thereby to preserve the experience of witnessing the sublime object and communicate that experience to others, began long prior to its completion. But due to the difficulty of accessing its remote location, the typical experience of the dam site for the vast majority of people has come not through witnessing its physical reality, but as an imagined experience through *mediated* discursive interactions of verbal and especially visual representation.

Boulder Dam ascended to its reverential status on par with that of the Golden Gate Bridge or Niagara Falls through its documentation portraying such notions as efficiency, modernity, masculinity, bounty, Manifest Destiny, and domination of nature. Indeed, the capacity of the products of human invention to touch the sublime through technology (often mediated by documentary technologies) emerged as competing with or even superior to that of nature's wonders, especially the technologies developed in the wake of Western industrialization and the heretofore-impossible structures they made possible. This so-called technological sublime exists in relation to the cultural context of technology as one of America's defining ideas about itself, one that has worked as a communal force binding together a multicultural society.[23] Theorists of the technological sublime argue, in fact, that in pluralistic and multicultural America, technology has allowed the creation of communal experiences that have replaced religious observances and function as sites of public display and cultural confluence that have both political and social signification for many Americans.[24]

In Hoover Dam we find advanced engineering techniques and newly devised machinery domesticating the initially terrifying canyon and river, replacing the natural sublime with a human-built one: a landscape created by nature was replaced by an engineered landscape created by human technology. This is not to say that the technological sublime completely usurped the natural sublime; sublime terror, an integral feature of Burke's natural sublime, was a day-to-day experience for those working on the dam. Theo White expressed this sentiment in a June 1935 *Harper's* article this way: "It is a beautiful tantalizing thing. It is complex. It has meaning, not to be grasped in weeks or perhaps years. It is subtle, sometimes cruelly obvious. . . . I stay on, fascinated. I stare at the thing trying to comprehend it, to fix it forever in mind's [sic] eye. I have been inspired and provoked in the weeks I have tried to know it. And now, the minds that invented it, the bodies that are building it, the complexity and spirit of it, the love of it which men feel—all, all of it bewilders me."[25]

Gigantic is a term that was often used to describe the dam, and is one that is best understood relationally; our basic relation to the gigantic (object) can be thought of in the same way as our relation to nature. Our immediate connection to nature or the environment is that it surrounds and envelops us. We cannot disassociate ourselves from nature to become an objective onlooker because we are within nature, and one could go so far as to say that we *are* nature. According to Susan Stewart, ours is a philosophical limit-problem in that any viewer of the gigantic cannot cease to be part of the subject; one cannot fully step outside of the gigantic (or nature) and thus cannot take full measure of its total expanse and influence. Just as we are enveloped by nature, the gigantic encloses us within its shadow, yet often without our awareness. Stewart defines its opposite term, *miniature*, as the depiction of an object's reduced physical dimensions and qualities on a scale so reduced as to suggest a world not necessarily known through our lived experience. She uses children as an appropriate metaphor not simply because they are "miniature" adults, but because "the world of childhood, [is] limited in physical scope, yet fantastic in its content . . . [yet] remote from the presentness

of adult life. We imagine childhood as if it were at the other end of a tunnel
—distanced, diminutive and clearly framed."[26]

Similarly, we can think of sublimity in terms of both gigantic and miniature. Derrida challenged Kant's mathematical sublime (a sublime so inesti-
mable that humans could only have access to it through mental abstraction)
with a notion of infinitesimal sublime: if the impossibly large is sublime,
then so too is the impossibly small, a conception that provides an opening to the difference between sublimity and other forms of human experience.[27] He considers that the answer lies in proper perspective of the sublime, relying on the originary sense of the term, which means "elevation"
or "grandeur," concluding that it would be silly to consider the greatness
of writing and oratory as being physically larger. Instead, we can imagine
it to be greater, loftier, elevated, and astonishing. Thus, it is not the actual
physical size that evokes sublimity, but the relative evaluation of the object
or event, the dynamism of the scope of contrast, and the scale of the *relationship* between the object or event and ourselves that make everything else
(including ourselves) seem insignificant in comparison.[28]

The Penstock Pipes

Some of the most visually impressive aspects of Hoover Dam's construction
—for which the relationship between gigantic technology and tiny human
figures is a main feature of the object's presentation—were the penstock
pipes, the fusion-welded plate-steel pipes designed to take water from
the intake towers to the powerhouse and outlet works below the dam. Too
large to transport over any significant distance, the pipes were fabricated
at a facility built near the dam site and driven only a few miles to the dam
on special tractors made specifically for this project. A 1932 booklet by the
Babcock & Wilcox Company titled, simply, *Hoover Dam* explains their role
in making pipes that were "larger than any heretofore made," and only possible through the newly discovered process of examining the strength of
welded seams by X-ray equipment.

The pipes were a favorite subject of photographers during construction, and their massiveness was conveyed by interanimating the miniature
human and the gigantic pipe. Babcock & Wilcox manufactured different-
sized pipes—diameters of eight and a half, thirteen, twenty-five, and thirty
feet—depending on the services for which they were intended, yet nearly
every photograph is of the largest pieces. By means of the photographed

contrast between human and engineered product, the pipes could be seen as a microcosm of the engineered sublimity of the entire project in which the effect of the sublime structure is to make everything else (particularly people) seem tiny and insignificant in comparison. But more importantly, since the penstock pipes were to be buried within the canyon walls and would be inaccessible and visually absent once in place, photographs of the pipes in fabrication or en route were the only chance to chronicle their existence and celebrate their colossal size.

The sentiments of miniature and gigantic are expressed through what I have referred to as a visual parataxis, in this case the recurring motif in which the impossibly large penstock pipes are set in striking juxtaposition to the relative insignificance of the human figure. In many instances we see men standing inside a section of pipe in which they appear minuscule compared to the gigantic cylinder, yet stand proudly and obviously posing for the photograph. Unlike many of the images taken at the dam site that appear to be "candid" photos of laborers going about their work and unaware of the photographer, images of the penstocks are nearly always posed pictures, as if those posing (and photographing) saw these behemoths as monumental in their own right and thus candidates for commemoration. These are portraits of glorious achievement and a visualization of how the advance of human engineering has enabled men to create such sublime objects. Babcock & Wilcox, in fact, proudly deployed this technique in its own promotional materials, such as in a 1933 booklet also titled *Hoover Dam*, which shows the pipes in relation to people in a variety of illustrations and photographs. American Steel & Wire, in their 1931 booklet *White Gold: From the Turbulent Waters of the Colorado*, also illustrated the penstock pipes with small human figures for comparison. The Bureau of Reclamation and the Los Angeles Department of Water and Power also produced numerous photographs of people next to or inside of the giant conduits.

In figure 4.02 we see a composition of technological sublime enhanced by the illusion of danger within the natural sublime. This photograph, of which there are several extant variations from the same photo shoot, is framed such that the pipe appears so large that the camera frame can only capture half of it, yet the pipe also appears to be hanging high above the canyon with the dam far off in the background. The image causes us to wonder how such a massive piece of metal, one so large as to easily accommodate twelve men and seemingly with room for fifty more, can be suspended so high above the canyon floor. In the photograph, several members of the Boulder Dam Consulting Board, the Bureau of Reclamation, and the Babcock & Wilcox Company are shown calmly posing inside the pipe. All of the men are wearing attire appropriate to engineers and executives—suits, ties, and hats— and all appear at ease as some look at the camera and others look out at the canyon, all postures suggesting that the men must be exceedingly brave to so coolly stand inside this dangling monstrosity hundreds of feet above the canyon floor posing for photographs while enjoying the view at the same time. As with other similar images, this one is not a candid or "documentary" photograph of normal dam construction events, but a staged image, or, as one journalist put it, a manufactured "stunt," to use as publicity material for the Bureau of Reclamation.

Figure 4.02. "Members of Boulder Dam Consulting Board and Officials of the Bureau of Reclamation and the Babcock and Wilcox Company in a Thirty-Foot Diameter Penstock Pipe Section Over Boulder Dam Power Plant," January 15, 1935 (photograph, Ben Glaha, Bureau of Reclamation). Courtesy of the US National Archives and Records Administration.

Figure 4.03. "Train Through Penstock Pipe," March 13, 1934 (photograph, Bureau of Reclamation). Courtesy of the Boulder City Museum and Historical Association.

One of the more ambitious attempts to showcase the enormity of the penstock pipes is a photograph that features a threefold reduction in size. In figure 4.03 we see a publicity photo in which the Babcock & Wilcox Company secured a piece of piping under railroad tracks, then positioned a train inside the pipe, and then positioned human figures standing next to the train. There are extant photos that show several angles of this scene, one of which appeared on a souvenir postcard, while the Boulder City Museum and Historical Association later had fiftieth-, sixtieth-, and seventieth-anniversary commemorative coins struck picturing the train inside the pipe, providing further endorsement of the celebrated sublimity of the dam's manufactured parts.

The Incomparable Dam

From the moment of its completion Hoover Dam was placed on the short list of the world's great wonders. In an article that appeared in the December 1935 edition of the the *Rainbow* (Nevada), R. Robert Russell asserted, "Boulder Dam should be visited by every American because it stands alone, monumental, beyond all other world attractions." The article begins, "On the Colorado River, near Las Vegas, an army of men and a vast array of giant machinery are completing one of the greatest engineering projects the world has ever known—the Boulder Dam. It is so colossal it staggers the imagination. For engineering ingenuity, co-ordination of construction activities, together with the unheard quantities of necessary material, it is almost beyond the comprehension of the layman."[29]

Here Russell isolates the inherent conflict in our relationship to the sublime: how does one describe something that is "so colossal it staggers the imagination" or visualize what is "almost beyond comprehension of the layman"? Because of the unprecedented scope and scale of the project, many writers were confronted with the problem of animating for a reader what they had witnessed without devolving into empty hyperbole. Florence Lee Jones described this in a 1946 retrospective newspaper article, writing that "practically every piece of major equipment used was the 'biggest of its size ever produced'" and was one-of-a-kind technology, made uniquely for this undertaking.[30]

Newspaper publishers addressed the challenge by regularly evoking the vocabulary of the sublime in their headlines, journalists by writing superlative-filled descriptions of what they saw, and photographers by framing the magnitude of the dam or its elements against the comparatively miniature size of human bodies. Often the "you just can't believe how big this is" accounts came in the form of both statistical descriptions and contrast to things with which typical readers might already be familiar. For example, Russell writes, "The Dam will develop one million eight hundred seventy-five thousand horsepower, or four times the power developed on the American side of Niagara Falls. This is more than the power development of Niagara Falls, Muscle Shoals, and Dnieperstroy combined."[31] Russell continues,

Five and one-half million barrels of cement were used on this project. This is seven hundred and fifty thousand barrels more than has been used by the Department of the Interior in twenty-seven years of gigantic construction. Such a quantity would make a standard roadway sixteen feet wide from Miami to Seattle. Its weight is 7,300,000 tons, a mass larger than the Great Pyramid. The dam is 752 feet high, almost twice as high as the Los Angeles City Hall, and more than twice as high as the Mormon Temple in Salt Lake City. . . . Boulder Dam will create the largest artificial lake in the world. . . . Enough water will be impounded behind the dam when filled to supply five thousand gallons for every man, woman, and child in the world. If a billion gallons were to be drawn daily from the lake . . . it would take twenty-nine years to empty it. This volume would cover the state of Connecticut to a depth of ten feet.[32]

Here, in an attempt to provide abstracted mental references of comparison for the reader, Russell likens the dam project not only to Niagara Falls but also to continent-length highways, the Great Pyramid, Los Angeles City Hall, the Mormon Temple, and lakes the size of Connecticut.

Dorothy Childs Hogner attempted to describe her first visit to the dam in similar terms. She writes, "When we begin to think of figures in connection with Boulder Dam, we grow dizzy. Nine million tons of rock was excavated, and over a thousand miles of steel pipe were installed. Five million barrels of cement were used, and ninety-six million pounds of metal. Does that give you some idea of the scale upon which things were counted? The immensity of the project quite bewildered us."[33]

Although statistical comparisons, such as those Russell and Hogner offer, appeared with great frequency in newspapers, magazines, and government promotional materials, visual comparisons were also abundant. Although I noted previously the paratactic presentations of the Washington Monument and other objects superimposed on images or photographs of the dam, people posing inside the penstock pipes, and so forth, this technique became, if anything, *more* prevalent in the dam's post-construction era. One poignant example is in the 1937 Boulder Dam Service Bureau's *Boulder Dam Book of Comparisons*, a boldly colored booklet that depicts the dam verbally and visually in relation to everything from the pyramids of Egypt to locomotives, the *Queen Mary*, and the world's tallest skyscrapers.

It is not known if the authors of the *Boulder Dam Book of Comparisons* read Russell's account, but one page replicates his comparisons. Figure 4.04 shows a page on which we see a row of locomotive engines along with renderings of Niagara Falls (a comparison I discuss at length later in this chapter), Muscle Shoals, and Dnieperstroy (spelled "Dnieprostroy") Dam. As the caption in the middle of the page indicates, the power-generating capacity of the Hoover Dam is more than that of those three celebrated waterworks combined. On another page we see the weight of the dam compared to the North American continent, a row of *Queen Mary* cruise liners, and a stream of pilgrims carrying one-hundred-pound blocks. On yet another page we find the dam compared to several "famous tall buildings" of the time, a theme seen in many representations of the dam. In this image, the dam shares the visual field with the Washington Monument, Los Angeles City Hall, the Chicago Board of Trade, the Eiffel Tower, and the Empire State Building. The dam towers over all of these buildings (except the Eiffel Tower, which is set to the side and blends with the background), and its combination of width and height dwarfs the other structures on the page. The only building

Figure 4.04. "Locomotives, Niagara, Muscle Shoals, and Dnieprostroy," 1937 (in *Boulder Dam Book of Comparisons*, Boulder Dam Service Bureau). Courtesy of the Boulder City Museum and Historical Association.

significantly taller than the dam is the Empire State Building, which is conveniently—and deceptively—placed on the next page. Although the Empire
State Building is almost twice the (actual) height of the dam, it is not drawn
to scale as the other buildings seem to be and so appears only slightly taller
than the dam, thus preserving the dam's visual domination of the scene.

Another example of this strategy is shown in the 1931 booklet *White Gold:
The Story of Hoover Dam,* from the American Steel & Wire Company of New
Jersey (figure 4.05). The booklet traces the history of the American West
and its relationship to the Colorado River and shows the dam towering
over a skyscraper with the caption "And the tallest building in the far west
is dwarfed by comparison." Here the paratactic relationship is continued
by placing the dam in a dominating position over other constructed
technologies—in this case buildings—with which others might be familiar,
and just in case the readers are not, the text will provide the contextual cue.
We also see in this image clouds rising above the dam, as well as the winged
eagles of the original design, lending a sense of grandiosity to the entire
presentation.

Comparisons to the pyramids of Egypt were an especially popular commonplace. Sculptor Oskar J. W. Hansen, who was selected by architect Gordon Kauffman to design the ornamental elements for the dam project, went
even further than most newspaper headlines by writing that the dam "represents the building genius of America in the same sense as the Pyramids
represent that of ancient Egypt, the Acropolis that of classical Greece, the
Colosseum that of imperial Rome, and Chartres Cathedral that of the brood-
ing religious fervor which was Gothic Europe."[34] The Bureau of Reclamation
in its 1936 promotional booklet *Boulder Dam* states, "More massive than the
Pyramid of Cheops in Egypt, Boulder Dam was built by 4,000 men in four
years whereas the great Pyramid consumed in its construction the labors of
100,000 men for a generation."[35]

Hansen and others clearly thought of the dam as a historically monumental undertaking that would stand for eons. Hanson described the dam
as "imperishable against the oblivion imposed by time," an achievement
that would survive well beyond his lifetime and even all of humanity.[36] An
astronomer and mathematician as well as a sculptor, Hansen's star chart
embedded into the terrazzo floor of the dam's visitor's gallery measures in
"Platonic" years.[37] Designed to be accurate for the next fourteen thousand
years, it communicates to some future beings the great moments in human

AND THE TALLEST BUILDING IN THE FAR
WEST IS DWARFED BY COMPARISON

Figure 4.05. "And the Tallest Building in the far West is Dwarfed by Comparison," 1931 (in *White Gold: The Story of Hoover Dam,* American Steel & Wire Company of New Jersey). Courtesy of the Boulder City Museum and Historical Association.

history, including the birth of Jesus Christ, the building of the Pyramid of Cheops, and the exact date and time of Hoover Dam's dedication.[38] Such future beings would presumably be impressed by the millennia over which the dam has stood in place. Hansen writes, "Day by day, and in remote ages to come, intelligent people may view the star map out of which rises the monument at Boulder Dam and from it learn that the astronomical time of the dam's dedication was in the year 1935 of Our Era, on September 30, and at 8:56, 2.25 seconds in the evening of that day, as calculated from the center of our Sun, or the center of the Ecliptic."[39] Hansen's is a "plea to future peoples to carry on and build for future minds, knowledge concerning the true nature of our own and those other worlds in space."[40]

A Veritable Niagara

On September 11, 1936, the *Chillicothe Constitution-Tribune* of Chillicothe, Missouri, ran the front-page headline "10,000 Persons See Boulder Dam Valves Opened." The article expressed the sentiments of many who witnessed the opening of both sides of Hoover Dam's jet flow valves for their first-ever simultaneous test, one that inspired countless comparisons to Niagara Falls:

> Thousands of persons began arriving at the damsite during the morning hours to witness the ceremony. Civic officials congregated in the giant powerhouse where temporary grandstands had been erected for the event. . . . The mightiest waterfall ever harnessed by man . . . a demonstration that rivals Niagara Falls. While 10,000 spectators watched in awed silence all twelve outlet valves that keep waters of the Colorado river walled behind a $168,000,000 dam, began to open. Surpassing Niagara Falls in magnitude the water swept through the flood gates of the horseshoe-like dam met in mid-air and then crashed into the canyon below sending up a spray almost to the top of the 727 foot dam.

In some ways, the comparison of the dam to Niagara Falls seems appropriate. For much of early American history, Niagara Falls was the quintessence of American and natural sublimity.[41] From painters such as Thomas Cole (*Distant View of Niagara Falls*, 1930) and Frederick Edwin Church (*Niagara Falls*, 1857) to writers such as Charles Dickens, James Fennimore Cooper, Margaret Fuller, and Mark Twain, Niagara Falls inspired artists and writers to attempt to express their awe through prose and images.[42] Travel guidebooks, advertisements, songs, stories, paintings, and word of mouth brought visitors from around the world to view the spectacle in person.[43] The same can be said of Hoover Dam, as writers, politicians, and artists lauded it as a masterwork of engineering, and travelers attempted to express their experience in letters, photographs, and other testimonials of witnessing the dam.

Niagara Falls served as a compelling analogy for Hoover Dam's jet valve tests.[44] The April 22, 1936, edition of the *Boulder City (NV) Evening Journal* featured an image of an April 4, 1936, valve test with the title "Boulder Dam's Niagara" and a caption that begins, "Here's Boulder Dam's Niagara Falls in action—half of it" calling the sight of the cascading water a "sensational result." The caption goes on to preview the full May 4, 1936, test as a moment in which "water from the two falls will battle for a place within the canyon." The caption also notes that the volume of water will be "approximately the amount flowing over Niagara." A 1939 booklet by the Boulder Dam Service Bureau titled *Boulder Dam and the Boulder Dam Tourist Recreation Area* includes a colorized illustration of an Arizona side valve test with a caption that reads "Higher than mighty Niagara Falls is the water discharge from the canyon wall." A 1939 *Popular Mechanics* article states, "At Boulder dam there are more than 21,500,000 pounds of gates and valves, a few weighing as much as 500,000 pounds. Opening such values releases a veritable Niagara under pressure of hundreds or even thousands of pounds."

It might be helpful here to explain some terminology for different functions of the water bypass systems at the dam that allowed this "Niagara" to occur. The spillways are designed to route water around the dam during flood conditions. They are concrete-lined, open channels about 650 feet long, 150 feet wide, and 170 feet deep, and constructed upstream of the dam on each side of the canyon.[45] The jet flow outlet works are downstream of the dam and include four jet flow gates in the inner diversion tunnels—two jet flow gates each in the Arizona and Nevada canyon wall outlet works (also known as the valve houses) located in the canyon walls about 180 feet above the river. These gates are also designed to route water around the dam under emergency or flood conditions, as well as to empty the penstocks for maintenance work.

Although the spillways had only been tested one time (on August 6, 1941) and the jet flow outlet valves were tested fewer than a dozen times in the five years between 1936 and 1941, images of the dramatic scene when one or both of the valves were opened—the only dramatic event amid otherwise pedestrian, even tranquil, day-to-day activities at the dam—are by far the most numerous (and memorable) visual representations of the dam during this time period.[46] Newspaper articles, brochures, and myriad other publicity materials featured images with the valves open. Even before construction began, dramatic illustrations depicting water cascading out of the canyon sidewalls were abundant. Figure 4.06 shows one of the first architectural renderings of the dam, which depicts two separate outlet works on the Arizona side (but none on the Nevada side) pouring water into the downstream river. This image was reproduced in numerous publications, with just two examples being the 1931 *Popular Mechanics* article "A Glimpse into the Future at Boulder Dam," as well as on the front cover of the January 15, 1932, *Las Vegas Evening Review-Journal,* with a nearly full-page reproduction of the illustration.

The practice of using the more dramatic, yet rarified valve test rather than the commonplace scene as the representative presentation of the dam continued throughout construction and beyond. *Electric West* magazine used a full-page cover photograph of a valve test in their October 1936 *Boulder Dam Power* Pictorial History edition. The cover of the 1937 *Boulder Dam Book*

Figure 4.06. Updated early rendering of Hoover Dam. Courtesy of the Nevada State Museum.

of Comparisons featured not a comparison, but an illustration of a full valve test. The August 1937 edition of *Travel* magazine includes a photo spread with the Nevada side valves open, while the accompanying article features additional aerial photographs of Hoover Dam with claims that it rivals the most revered naturally sublime place in America—the Grand Canyon.

Some images combined the drama of the valve tests with the sublimity of the nighttime dam site discussed earlier in the chapter. Recalling a 1935 Bureau of Reclamation photo of the outlet works illuminated at night, the Boulder Dam Service Bureau's 1938 booklet *Pictorial Boulder Dam*—which, it says, "ignore[s] the more material aspects of Boulder Dam. . . . [F]acts and figures must be forgotten for the moment in favor of the soul-satisfying esthetic pictorial appeal" (ellipses in original)—includes as its cover image a hybridized nighttime dam photo and illustrated stream of water exiting the Nevada side jet flow aperture, while an image inside the booklet shows the 1936 full jet flow valve test. The 1939 Boulder Dam Service Bureau's *Boulder Dam and the Boulder Dam Tourist Recreation Area* booklet includes not one but three different illustrations of the dam with its valves open. One image shows the Arizona side open, a second shows both sides open, while a third is similar to the 1935 and 1938 images' nighttime scenes. This third image is the only one of the three to include an extended caption, which describes the nighttime scene as "inspiring" and contrasts it with a daytime one, in which the dramatic landscape distracts visitors from the dam's magnificence. The caption of the nighttime scene continues, "The cliffs of Black Canyon, darkened to unimportant dullness, retire into the shadows, for the Dam to display without competition from Nature the tremendous features which render it 'Man's Greatest Engineering Achievement.'"

In an image I will return to later, the cover of the November 1939 *Country Gentleman* magazine shows three men watching a Nevada side valve test, while the June 7, 1941, *Boulder City Journal* features a circular photograph of a full valve test with the caption "Boulder Dam, Southwest's Greatest Power Project and Tourist Attraction." A Bureau of Reclamation photograph of the full valve test also graces the cover of the January 16, 1944, *Arizona Daily Star*.[47] *Arizona Highways* magazine also ran several valve test images of the dam: the July 1941 edition includes an image with both sides open; the April 1950 edition presents a full-page color photograph of the dam with both sides open, complete with a rainbow in the foreground; and the March 1956 edition features a full-page cover photo with the Arizona side valves open.

The "Niagara" experience was not just offered in images, however. Similar to the boat rides offered at Niagara Falls, the Bureau of Reclamation also offered "rides" to allow certain dignitaries to experience the outlet water release up close. The February 1950 edition of the *Desert Magazine* included a photograph of people on a platform suspended high above the river by a cableway, the same cableway that was used to deliver laborers to the canyon floor and deposit the millions of tons of concrete for the dam (figure 4.07).[48] In this photo, originally taken by Cliff Segerblom for the Bureau of Reclamation after heavy rains raised Lake Mead to record levels, we see water pouring through the jet flow valves on each side of the canyon while the tiny spectators dangle precariously close to the streams.[49] Segerblom described the circumstances surrounding this shot in the May/June 1985 edition of

Figure 4.07. "Canyon Wall Outlet Works," September 28, 1940 (photograph, Cliff Sergeblom, Bureau of Reclamation, later reprinted in the *Desert Magazine*, February 1950). Courtesy of the Boulder City Museum and Historical Association.

Nevada: Magazine of the Real West: "I understand it was the only time all the outlets were open at once. At one point I was on that skip, which was carrying a number of dignitaries. Interior secretary Ickes was on it, as I recall. I took some photos and then got off and ran around on top of the dam."

From the extensive circulation of open-valve images, one might get the impression that the "Niagara" waterfall at the dam is an everyday occurrence, an insight surely not lost on promoters of vacations and travel to the dam site. The dramatic images of water bursting out of the side of the canyon into a gigantic cloud of mist provided a pervasive, albeit misleading, notion of what travelers might see when visiting the dam. Although none of these individual images are *the* iconic image of Hoover Dam, the frequency with which this spectacle appeared in dam-related imagery—the aggregate of these repeated images over time—became the way in which the public likely imagined or envisioned the dam. The selected images of the jet-flow test scene are in some ways similar to the imaginative illustrations of the project in the pre-construction phase. As Paul Ricoeur points out, although our term *images* can refer to representations of existing but absent things, it can also refer to *illusions* and to deceptive representations of the nonexistent.[50] Although the photographs of the jet valve tests are not illusions, they *are* deceptive representations: snapshots that freeze and preserve a moment of a seldom-occurring event, the promulgation of which conveys the false impression that a visitor to the dam will actually witness the spectacle.

From the Sublime to the Picturesque

Paul Messaris writes, "As van Eck points out, one of the most central lessons that Renaissance artists drew from classical rhetoric was the importance of establishing 'common ground' between the speaker and the audience. By extension, pictorial perspective was considered a means of creating the illusion of a literal common ground—a space that the viewer might enter."[51] In Caspar David Friedrich's well-known canvas, *Wanderer Above the Sea of Fog* (1818) (figure 4.08), we see a motif that originated in the Renaissance but is particularly associated with nineteenth-century landscape painting— the *Rückenfigur,* a person seen from behind contemplating the scene before them. When an individual faces the viewer in a painting or photograph, he/ she often overpowers the surrounding environment by drawing our attention away from the proximate circumstances, instead focusing our attention on the persons in the image. But *Rückenfigur* vicariously situates the viewer

Figure 4.08. Wanderer Above the Sea of Fog, 1818 (oil on canvas, Caspar David Friedrich).

as the individual, as the prospective traveler or the spectator of the scene.[52] The individual in the image directs the viewer's gaze beyond themselves and invites the viewer to enter the pictorial frame and participate in the depicted experience.

Friedrich's painting depicts man's contemplation of nature. The wanderer happens upon an awe-inspiring view and is presented with a vision of sublimity. The wanderer surveys the land before him; however, we cannot see his face, and thus it is impossible to know whether he experiences the

Figure 4.09. "Looking downstream from the same point as No. 989," March 14, 1932 (photograph, Bureau of Reclamation). Courtesy of the University of Nevada, Las Vegas Libraries, Special Collections. W. A. Bechtel Collection.

landscape as exhilarating, terrifying, both, or neither. The staging invites a two-sided interpretation as the image conveys the wanderer's mastery over the landscape, as well as his insignificance within or relative to it.

The *Rückenfigur* is also a prominent motif in Hoover Dam imagery. The device is used in numerous images in which figures are shown on a precipice looking down to the Colorado River, Lake Mead, or the dam itself. One of the first examples of this appears in a March 1932 Bureau of Reclamation photograph that shows the lower portals of the diversion tunnels being excavated (figure 4.09). In the image we see a downstream view of a man sitting on a rock ledge overlooking the river. Far below we can see construction activity on both sides of the river, which the caption of the photograph describes as a "graphic view of the problem of muck disposal." Here we see a man in the lower foreground of the frame, but because his back is to the camera and he is gazing on the scene below, the composition invites us to place ourselves in the man's position to contemplate the scene within the canyon just as we envision he is doing. Yet, because his posture seems relaxed—he is sitting down, arms across his knees—we get the feeling of serene contemplation, not fear or anxiety.

The photograph of the dam site most reminiscent of Friedrich's painting is from an August 1932 image, the framing and composition of which call to mind Friedrich's work in several ways (figure 4.10). In both the photo and the painting we see human figures (the wanderer in the painting and high-scalers in the photograph) standing on a small, rocky outcrop gazing down at the landscape below. In neither image can we see the figures' faces; they are obscured, which conceals their emotions and reactions to the scene. Finally, in both images, we see cliffs rising up to peaks in the background, but the haze obscures the details of the distant landscape. Just as in the painting, there is a tension between sublimity (terror) and safety in the photograph. Although the men in the photograph are at great height, they appear to be relaxed, with one man going so far as to extend a leg out into midair. But unlike the wanderer who stands on the precipice unprotected, the high-scalers are enveloped in safety equipment: hard hats, lanyards, ropes, gloves, boots, and so forth. The caption works to dispel the sense of safety, however, and to increase the dramatic effect of the image by indicating to the viewer that the high-scalers are "over 1000 feet above the river."

Despite their similarities, the *Rückenfigur* motif of a photograph works in ways that the painting does not. We bring to the medium of the canvas an expectation that it is an interpretive visualization of a scene and can be painted from any viewpoint or angle the artist desires. Conversely, a human photographer takes a photograph, and the viewpoint of the image is also the viewpoint from which the photographer (at least in this time period) must be physically positioned. A well-known example of this is Lewis Hine's photographs of workers on the Empire State Building, which were remarkable in part because Hine himself had to scale the building to occupy the same space as those he was photographing. The viewpoint of the photographs

Figure 4.10. "High-Scalers perched on a bench over 1000 feet above the river prepare anchorages from which workmen will be suspended by ropes in scaling the canyon walls for the penstock tunnel audits," August 22, 1932 (photograph, Bureau of Reclamation). Courtesy of the Special Collections, University Libraries, University of Nevada, Las Vegas. W. A. Bechtel Collection.

provided a seemingly God's-eye portrayal of the construction process. Likewise, the 1932 high-scalers photograph (and every subsequent photograph using the *Rückenfigur* motif) encourages viewers to place themselves not only in the position of the person(s) depicted but also in the position of the person doing the depicting.

Figure 4.11. "Boulder Dam and Power Plant with discharge from two 84-inch needle valves as seen from high mountain down stream from dam. Fortification Mountain in the background," April 14, 1938 (photograph, Bureau of Reclamation, later reprinted in *Country Gentleman,* November 1939). Courtesy of the US Library of Congress, Prints and Photographs Division, Washington, DC.

A photograph originally taken by Bureau of Reclamation photographer Ben Glaha (figure 4.11) that later appeared in the November 1939 issue of *Country Gentleman* also brings to mind Friedrich's use of the *Rückenfigur.*[53] In it, we see three men, all from behind, gazing down upon the dam during a valve test. The open valves discharge water out into the canyon in front of the stark white dam wedged between the cliffs as tall mountains rise behind the dam. Bright white clouds fill the sky, while clouds not seen in the image cast shadows on the canyon walls, enhancing the contrast of dark and light.

Although, as with Friedrich's wanderer, we cannot see their faces, the men appear calm as they take in the powerful aesthetic experience. One man stands with his legs slightly spread and hands in his pockets, while the other two sit relaxing on the cliffs. Like Friedrich's wanderer, the individuals in this image have stopped to contemplate a magnificent scene; however, in this case they enjoy a view that fuses natural and technological sublimity into a picturesque scene. Although the image invites viewers to imagine the people in the image as their surrogates or guides, this photograph implies an additional inference: that the experience is not terrifying, but attainable. It is the latter aspect that makes the *Rückenfigur* an especially powerful photographic device for the promotion of tourism.[54]

Another Bureau of Reclamation photograph, one later reprinted in the July 1941 edition of *Arizona Highways*, a magazine specifically devoted to promoting tourism in the state, shows a woman (the photographer's wife) sitting, looking down at the top of the dam (figure 4.12). The woman is seated on a sheer cliff, yet she appears to be quite relaxed, sitting with one leg out and the other casually folded underneath her. The image, however, stands in stark contrast to the language used in the accompanying article. Although the scene in the photo is tranquil, the text describes the dam—"a powerful servant of the American people"—as something built to "harness the mad Colorado and to translate the fury of the flood into power for man's use." An accompanying image of a valve test on the same page (the same September 28, 1940, image taken by Cliff Segerblom of dignitaries taking "rides" just above the gushing waters, discussed earlier) depicts what might be described as "fury," yet that fury was man-made and photographed specifically for its spectacle. Conversely, the text goes on to describe the now-submissive Lake Mead as "blue and calm" and "bathed in sunlight" and "a giant resting between high cliffs ready to do a giant's work," while another photo shows the turbines inside the dam and notes, "When a river is properly harnessed, the river works for man."

The "general view" photograph, however, represents a major shift in the visual discourse surrounding the dam. Because nearly every image associated with the dam's construction up to the late 1930s featured men, while the verbal descriptions regularly invoked the terror of Burke's sublime, the fact that a woman is able to sit atop a sheer cliff to command a view of the structure suggests that the danger previously associated with the project has been eliminated; thus, this image visually achieves a

Figure 4.12. "General View, Gene Segerblom Foreground," 1940 (photograph, Cliff Sergeblom, Bureau of Reclamation, later reprinted in "Power: A Pictorial Study," *Arizona Highways*, July 1941). Courtesy of the Boulder City Museum and Historical Association.

different type of *Rückenfigur* insinuation sought by promoters of tourism to Hoover Dam: if it is safe enough for a casually dressed woman sightseeing at the dam to get this vantage, it is safe for anybody. Gone is the natural sublime that the wanderer contemplated, replaced with a tranquil, domesticated, picturesque scene that stands in sharp juxtaposition to the images of men shown surrounded by protective clothing and safety equipment. That the woman sits in a dominant position looking down on various technologies—the dam, cars, trucks, roads, and so forth—further signals that the utter domination of the natural landscape is now complete.

In sum, this image presents the dam in a new guise—that of the picturesque. Although the sublime is often described as unframeable because it is boundless, in the picturesque the landscape as a perception of the natural world is reconstituted to meet human needs and changes to match our lived experiences.[55] The dam began as a sublime undertaking, one that was colossal and dangerous, inflicting death on man during its construction, and itself a process of man's attempt to prevent death from the river's floods. However, the picturesque encloses the sublime and constrains the extremes of nature. Instead of being terrified, we are imaginatively engaged and emotionally captivated.[56] In the picturesque, the terrors of natural sublimity are destabilized by its presentation in acceptable and digestible cultural forms that we find wonderful and enticing, but not limitless or terrifying.[57] In these images, the picturesque aesthetic maintains the symbolic legend of the sublime dam, but with what Judith Bernake calls "comfort-producing restraints."[58] This makes both the natural and technological sublime comprehensible and accessible, and therefore easier to accept.

The May–October 1955 edition of *Nevada Highways and Parks* contains one of several iterative photographs shot within a few minutes of each other (though we do not know in what order the images were taken), all comparable to the previous one, but which take the picturesque even one step further. In one image we see two women sitting on a precipice overlooking the dam. Below them is the same scene of the top of the dam with the lake to the left, cars and trucks traversing the dam, and so forth. Depicted in this image, however, is not the individual's contemplation of a sublime landscape as portrayed by Friedrich, but an experience shared between two people. But unlike the Glaha photograph, which centralizes the dam as one part of the broader natural landscape, this image barely acknowledges the surrounding terrain. Instead, the dam itself, with its stark whiteness and modernist curvilinear form, is the emphasis. If the sublime cannot be framed in words or pictures, then this image implies that the dam and landscape can at least be "realized" as part of a shared experience between people. Moreover, here we have a scene of intimacy, tranquility, and comfort quite at odds with any traditional conception of the sublime as dangerous, unruly, or awe striking.

Subsequent images continue to move (literally and figuratively) the *Rückenfigur* and consequently the viewer of the images away from the sublime and into the picturesque. Images in the March 1956 edition of *Arizona Highways* and again in the May 1964 edition depict the viewer at increasing distances from the dam site. In the 1956 image, the *Rückenfigur* is positioned on a precipice somewhere along Lake Mead with the dam far off in the distance. In the 1964 image, the pictured viewer is similarly positioned, yet the dam itself is not even visible. In 1959, *Life* magazine featured several full-page images of a couple taking an aerial tour of the dam in a helicopter while being photographed from another helicopter somewhat above them. We only know that it is a couple because of the accompanying captions telling us so, but here we are even further removed from the dam, yet asked to view its exquisiteness through the perspective of not just the photographer, but the photographer photographing someone else gazing upon the site.

The development of this visual trope is carried further yet in other subsequent depictions, such as in the 1976 Bureau of Reclamation booklet *The Story of Hoover Dam*, in which the part of the *Rückenfigur* is no longer played by a photographed person, but by a black-and-white sketched illustration of the men from the 1938 Ben Glaha photograph that appeared in *Country Gentleman* superimposed on a different image of the dam from nearly the exact same vantage. In this image, the sketched men look out onto the same technological and natural sublime, but only under close scrutiny does one recognize that this image bears a striking resemblance to the original Glaha but has no valve test discharge; thus, it draws on the intertextuality of its antecedent image but renders an even more tranquil scene than was originally depicted.

All of these images were used in the service of shaping a discourse about the dam and surrounding landscape as picturesque. Examined within a broader context of the sublime and picturesque, as I have attempted here, we can see how these images helped fashion the dam as safe, accessible, egalitarian, and welcoming. The iconographic nature of the images and the use of the *Rückenfigur* served to transform the dam from an awe-inspiring, dangerous object to a scenic, safe, and tranquil site for tourism.

IN 1930S AMERICA, large concrete and earthen dams were regarded as symbols of American ingenuity that showcased technological modernism and the values of efficiency, utilitarianism, and functionality.[59] The difficulty that writers and artists faced in depicting Hoover Dam to the public in words or images is inherent to the nature of the sublime and testifies to the dam's status as sublime object. Images of the dam attempted to approximate the sublimity of the dam in various ways: paratactic juxtaposition of tiny human figures and gigantic components of the dam, or iconic man-made structures and the even bigger dam; playing on the "reality" of documentary photography by framing the heroic engineer and colossal technologies in artistically staged and quasi-documentary forms; circulating dramatic spectacles of hydraulic testing that called to mind the natural sublime of Niagara Falls; and framing photographs that reference nineteenth-century landscape paintings of the sublime. As a result, the images in the public record sought to capture and glorify the dam as a sublime technological achievement. Whether through visual and verbal comparisons to human figures, pyramids, or Niagara Falls or by visual conventions reminiscent of the *Rückenfigur*, representations of the dam opened a rhetorical space for discourse about the sublime.

Although the sublime presents a paradox in which it is thought to be unframeable, indescribable, and unrepresentable, as we have seen, people have nonetheless remained undeterred in their attempts at descriptions and representations of sublimity of all sorts. If we consider the sublime as the psychological capacity for astonishment, even while accepting that words or photographs cannot replicate the bodily experience of "realizing" the sublime, we can see how images and words can do the work of impressing upon the mind of a viewer/reader a sense of the sublime by arresting his or her attention and forging a sense of the limits of the self in the face of grandiosity. As Jonathan Crary notes, when we are looking or looked at, we are normalized by embodied conditions of perception. When we are "hailed" or interpellated by an image, we need not see the "reality" of the object, but we can still be caught up in its visual spectacle. "Spectacular culture," he writes, "is not founded on the necessity of making a subject *see*, but rather on the strategies in which individuals are isolated, separated, and *inhabit time* as disempowered."[60] The closing off of discourse by the sublime—the failure of words or images to frame the unframeable or represent the unrepresentable —can, nonetheless, be rhetorically successful by constituting a testimonial embodiment of witnessing, by establishing a relationship between human-ity and nature, by presenting the conceivablity of the representationally inconceivable, by providing a discursive landscape through approximating the act of facing the sublime, and by constituting a relationship in which viewers come to appreciate their own finitude in the face of something spectacular.

5

The Unseen

Documentation of the Hoover Dam's construction is extensive, ranging from imaginative illustrations of some future dam to photographs of man and machine that chronicle the entire construction process. Although these images are an important constitutive element of the cultural mythology of the dam, there are two aspects of its visual record (and I would argue any visual record) that should be of concern to us, particularly when considering the unfolding of certain scopic regimes dominated by ideologies of the state: the absence of certain types of images, and the inability of any particular image to reveal "the truth" about a situation.

Although Roland Barthes posits that the camera provides a testimony of indexicality, or of what "really was there," he and other theorists have also outlined the many ways in which the camera also "lies." Not only does a photograph only document a single instance out of infinite possible instances, but cameras themselves were designed to mirror the ways in which Western cultures understand the rules of perspective, thus privileging a single, stationary vantage point in a wider field of vision and with greater clarity of focus across that field than the human eye can actually achieve. In other words, when you look at an object, your eyes bring that object into focus while everything surrounding it remains blurry. But with a photograph, the entire envelope of the image is equally in focus, whether in the center or at the edges. Moreover, the image we see as "realistic" is actually unrealistic in that the three dimensions we view with our stereoscopic eyes are rendered onto a two-dimensional medium, which is why we do not think that a photograph of our mother is actually our mother, but instead understand it as a flattened symbolic artifact of our mother.[1] In fact, Barthes himself famously describes the "truth" of a photograph—the idea that we perceive photographs to be an unmediated copy of the real world—in terms of "myth." Consequently, humans have developed technologies such as cameras that help us "see," but, in turn, our use of these technologies disciplines us to see in particular ways. In effect, we are engaged in a series of visual feedback loops in which our technologies of looking constrain our ability to conceive of the world differently and thus position us to further refine similar technologies of looking, thus further constraining our perception of the world.

Although we no longer ascribe the truth of photography in the same way that the positivists did, and few people today adhere to the notion that knowledge gained through empirical data is the only authentic knowledge, a photograph does in fact communicate some sense of what was there, however partial that may be. For us, images constitute a way of seeing and knowing the past by forming a kaleidoscopic menagerie of individual traces that embody a layered, albeit flawed, network of knowledge about "what was there."

Photographs take impressions of time and place and render those impressions with presence for the individuals viewing the photograph who are literally seeing the past in the viewer's present, characteristics that underscore one of photography's fundamental differences from verbal discourse. For example, although Barthes mourned his mother by writing about photographs of her, he really was literally looking at his dead mother. And although the "what was there" of a photograph is partial, and not "truth," photographs do reveal lived impressions of past experiences of human culture that are anchored in a specific place and time. In other words, there are irreducible certainties of a photograph's indexicality that are undeniable.[2]

The implications of this are immense when we think about the citizenry function of photography that scholars such as Ariella Azoulay ask us to consider.[3] Throughout the remainder of this chapter I discuss what I call the *unseen* images of Hoover Dam, and examine their effect on the events and people marginalized as a result of being excluded from imagery sponsored by the state. These considerations have important political and social implications, as some images were deliberately kept from circulation, others played on received stereotypes to reinforce the social standing of certain groups, while the distinct absence of still other types of images reflects the disciplining function that a lack of images can have.

Censorship in Bureau of Reclamation Images

The Bureau of Reclamation was still a feeble organization in the early 1920s, with little money, with little power, and in constant jeopardy of being eliminated. Established in 1902 with the assignment of developing irrigation in the West, the bureau had undertaken a series of smallish projects but was receiving minimal support and struggled to remain relevant. In 1914 Arthur Powell Davis, nephew of the famed explorer John Wesley Powell, became its director, and more than any other person it was Davis who championed the idea of a dam on the Colorado, one that propelled the bureau to national and international recognition and set in motion decades of dam building across the United States and throughout the world.[4] After Hoover Dam came Parker Dam, then Imperial Dam, then the All American Canal, then Davis Dam, Laguna Dam, Palo Verde Dam, Navajo Dam, Flaming Gorge Dam, Headgate Rock Dam, Bonneville Dam, Jones Valley Dam, and hundreds more throughout the United States, Asia, Africa, and South America. The construction of Hoover Dam saved the department and, in the words of Donald Worster, helped to make the Bureau of Reclamation the "most famous and accomplished desert conqueror in world history."[5]

It all started, however, with Hoover Dam, the bureau's big (and last) chance to make a name for itself. The drive for public support was paramount to the bureau's continued viability, and as a result it maintained a carefully orchestrated visual and verbal publicity campaign throughout the entirety of the project. The bureau tightly controlled its public image in large part through its control and direct distribution of construction photographs to the media. Newspapers, always clamoring for more action shots of the dam, had no need to incur the expense of sending their own photojournalists to the construction site since they could rely on a steady supply of images from the Bureau of Reclamation, which had its own full-time photographers permanently stationed on-site. This was an advantageous arrangement for the bureau in several ways. First, there was limited opportunity for "troublemakers" to take photos that did not fit the desired narrative, and the bureau did not have to expend much energy rooting out such undesirables. Second, the bureau could direct their photographers to produce images that expressly supported a particular narrative. Third, this arrangement allowed the bureau an opportunity to prescreen and thus tightly control the distribution of information, a procedure that was made demonstrably clear in a May 20, 1933, memo from Harold Ickes, secretary of the interior (and simultaneously director of the Public Works Administration): "No publicity matter with reference to the Department, or with reference to any individual in the Department in his official capacity, is to be given without the approval of the Executive Assistant in Charge of Publicity, or of the Secretary." The bureau went so far as to demand that other publications writing about the dam—public or private—submit a draft of their text for bureau review prior to its publication. The bureau's main concern was in restricting any information (visual or verbal) that might cast the project in a negative light. Although the Bureau of Reclamation published hundreds of images of the project in outlets of all sorts, it succeeded in keeping images that might have suggested any danger associated with the project out of public circulation. Although some images appeared in local newspapers showing wrecked cars or fires that consumed homes, images of construction problems or industrial mishaps (of which there were many) are practically nonexistent in newspaper articles or US federal government records.

The Bureau of Reclamation was acutely aware of its own public image, and as a result it maintained a policy to impede any criticism of the project by ushering forth a steady stream of images connoting efficiency, productivity, safety, and thrift. Although the majority of the funds for the project had been allocated before construction even began, it was nonetheless a massive expenditure of public money during the worst economic climate in the nation's history. The Bureau of Reclamation photographers also understood these limitations and worked within them. As Daniel Rosenberg describes it, Hoover Dam's construction was "as much an event of political theater as it was of engineering practice."[6] In fact, according to photographer Winthrop Davis, the start of construction was actually a staged photo op for the press set up by the bureau. "Well when they announced to the world that the construction started," Davis recalls, "we went down on the gravel bank and had all the press men and had *Pathay* and *Globe International* and went down there and [the bureau] put nineteen sticks of dynamite in the gravel bank and set it off and it made a big cloud of smoke and that was the start of construction of Boulder Dam. Nothing happened for a year later, but it was good advertising for newspapers and magazines, they ate it up."[7]

Davis had various encounters with both the bureau and local authorities that illustrate the lengths to which they would go to control both people and information. The first occurred when Davis arrived in Las Vegas with the idea of photographing the dam and its surrounding areas. After parking his car on the south side of Freemont Street, he wandered past several abandoned storefronts until he came across a hand-lettered sign that announced the headquarters of the Las Vegas Chamber of Commerce. Lifting the latch of the door, he found inside a disused and ramshackle room with a worn-out desk, a telephone, several rickety chairs, and papers strewn about the floor as the only indications of occupancy until he heard a toilet flush and a middle-aged man emerged from a bathroom at the rear of the building. Davis asked the man if there were any jobs in town, to which he responded that there were none and that no work would be done on the dam for over a year. The Chamber of Commerce representative seemed pleasant and struck up a conversation, during which he excused himself to make a phone call. Shortly after the "muffled" call, the chief of police arrived, alerted by the man on the phone that "vagrants" were about inquiring about work. The police chief demanded to see Davis's money and vehicle to prove he wasn't some kind of drifter, adding that there were "a lot of vagrants arriving in town" and that it was the standard policy of the police to "round them up, pack 'em a lunch, give them a water bag, and take them down to the California line and dump them."[8]

Other incidents involved Boulder City, an area controlled entirely by the Bureau of Reclamation and set up like a reservation with guards stationed at each entrance requiring citizens to show passes authorized by the bureau to gain access. Davis sought to document the pass requirements along with Six Companies' policy of issuing scrip for the workers instead of cash (the scrip was the only tender accepted for rent, food, water, gasoline, clothing, and other necessities), which he saw as a blatantly shameless example of the illegal graft that was part of Six Companies' culture and ignored by the government. One day he was caught attempting to photograph the scrip-counting operation, and his pass was revoked, which effectively terminated his ability to work at or freely travel about the dam site. Davis describes the situation this way:

> There was a lot of graft around here and I had my nose in it clear out around here. Especially when I first came in here and Six Companies were paying their men in scrip. A lot of people didn't know that down in the basement was all the equipment for counting scrip. I tried to take a picture of it, [but] they caught me on it.

Callison was their representative and that's when they started putting the screws on me. They thought I had more than I had. I didn't really get a picture of it but I wanted to.[9]

Davis petitioned Elwood Mead, who was either sympathetic to his pleas or was afraid that Davis had more information than he revealed, and so had his pass reinstated. Regardless of the reason, Davis was relieved. "If it hadn't been for the intercession of Elwood Mead getting my pass back," Davis recalled, "I would've been finished. The moral of the story is, don't get noisy when you see other people's pockets being picked, unless you have friends around."[10] According to Davis, it was not just in Boulder City that information was controlled. He recalls in various letters and interviews that he had to be vigilant when taking any photographs. "I had to move around with considerable caution," he remarked, "because contractors caught me in several places cause I was where I shouldn't have been. I had to be able to recognize the foreman to be able to get into some of these places."[11]

St. Francis Dam Disaster

The bureau was engaged in much self-promotion for its role in building the dam, but it was also presented with a particularly difficult balancing act for its publicity campaign: show heroism and ingenuity in building such an incomparable structure, yet at the same time convey to the public a complete sense of confidence that the largest dam ever constructed was actually safe. One reason for the bureau's uncompromising avoidance of references to danger and mishaps was likely a consequence of the St. Francis Dam disaster still being fresh in the minds of many people at the time.

Just a few years prior to the start of Hoover Dam's construction, the worst US peacetime catastrophe since the 1906 earthquake and fire of San Francisco occurred in the San Francisquito Canyon north of Los Angeles. Just before midnight on March 12, 1928, the St. Francis Dam, a 205-foot-high concrete gravity dam, began to collapse, sending twelve billion gallons of water rampaging through the Santa Clara River Valley, killing hundreds of people, leveling homes, and destroying livestock, farms, and orchards (figures 5.01 and 5.02).

Built under the direction of William Mulholland, the chief engineer for the City of Los Angeles and the architect of most of its water appropriation projects of the previous twenty years, the St. Francis Dam was initially intended to be much smaller. Mulholland, against the advice of his own

TOP: *Figure 5.01.* "St. Francis Dam before the 1928 failure," March 1928 (photograph, H. T. Stearns, US Department of the Interior, US Geological Survey).

BOTTOM: *Figure 5.02.* "Taken from the same location showing the remains of the St. Francis Dam and reservoir floor," March 17, 1928 (photograph, H. T. Stearns, US Department of the Interior, US Geological Survey).

engineers, decided that with the shortage of storable water and the continuing influx of people to Los Angeles, the dam needed to be able to store more water than had originally been planned.

But shortly after its completion, the dam began to leak. To reassure the public, Mulholland and his chief engineer went to the dam on the afternoon of March 12, 1928, to do a thorough inspection. Upon completing his assessment, Mulholland said that such leaking was typical of large concrete dams and thus declared the dam sound. Less than twelve hours later, the dam collapsed, leaving four hundred people dead, twelve hundred homes destroyed, and eight thousand acres of farmland obliterated.

Most of the water reclamation projects in the American West at the time were modeled on similarly huge gravity-type structures, but after the St. Francis Dam tragedy, the safety and stability of using concrete or earth fill to hold back such a considerable volume of water were cast into doubt. The *New York Times* reported that because of the St. Francis catastrophe, "attention is being centered on the dams and reservoirs throughout the country. In California and the West, of course, this interest is acute for a great many of the communities out there are faced with conditions almost identical to those of the country below the St. Francis Dam."[12]

In its effort to "make the desert bloom," the Bureau of Reclamation was building numerous other dams across the West, with several on the scale of St. Francis, which the *New York Times* reported was "now the subject of national debate, and is hedged about with a confusing mass of issues."[13] To add to the bureau's anxiety—in an irony no one could have foreseen—the St. Francis Dam failed on the eve of the passing of the Swing-Johnson appropriations bill in Washington.

For the bureau, the St. Francis incident portended the very real possibility of a public relations nightmare and perhaps a pause or even a reconsideration of Hoover Dam and other similar dams. A *Wall Street Journal* editorial suggested just such a scenario by arguing that the St. Francis Dam tragedy was an ominous warning for the need to curtail the project and a refutation of municipal ownership policies regarding dams and reservoirs in the West. However, the *Los Angeles Examiner*, always the stalwart champion of the project, quickly came to the defense of the bureau and condemned the editorial, calling it a "ghoulish and contemptible use of public calamity to serve private interests."[14] As a result of the St. Francis Dam collapse, the bureau sought to provide evidence to the public that Hoover Dam was, above all else, safe and permanent, an omnipresent dictum in both the visual and verbal presentations of the dam. Any reference to unsafe working conditions, construction problems, mechanical failures, or any other account that would suggest something other than a massive, indestructible, and impregnable dam might remind the public of the recent and epic failure of the St. Francis Dam and threaten to erode public and governmental support for any number of reclamation projects.[15]

The Appearance of Safety

As a consequence of the St. Francis incident, the appearance of safety and efficiency was of paramount concern for the bureau and Six Companies, a concern that was reflected in both the images that were circulated and, just as importantly, images that were not circulated. This point is most clearly evidenced in the fact that although thousands of photographs were taken of the dam's construction from start to finish, and well over one hundred people were killed while working on the dam, with scores more injured, no government images exist of maimed or dead workers, destroyed vehicles, or any other of the hundreds of incidents and accidents at the dam site.[16]

Although no images of industrial accidents appear in either Bureau of Reclamation records or national newspapers of the time, and very few incidents of any kind were published anywhere, there are many verbal descriptions of such incidents in oral histories and personal memoirs from those who lived and worked at the dam site. Winthrop Davis, for example, describes an accident with the cableway system that delivered supplies and concrete to the canyon floor as one of many that he witnessed. "Six Companies used to run their cables until they had fur on them," he said. "What I mean by fur on them, I mean the strands were broken on the outside."[17] According to Davis, Six Companies decided that it wasn't economical to take the time to fix frayed cables, so they would run them until they snapped, which sometimes resulted in thousands of pounds of cargo or concrete plunging hundreds of feet below to the canyon floor, injuring those unfortunate enough to be in the vicinity. In other instances, simple miscalculations caused severe injuries to workers. "I know one day the signal man got balled up," Davis recalled. "He was on the Arizona side and they had to stop that swing when it [was] just right and he miscalculated and [it] went right into the canyon wall, busted it into a lot of pieces and a lot of broken legs. I don't know if there was anybody killed that day or not. I heard the ambulance coming. I had a picture of the skip before it even started, but I didn't want to get mixed up in it."[18] Here, as on many other occasions that Davis recalls in his memoirs, he was concerned that if anyone suspected he had taken a photograph of the accident, he would be in danger of being cast out as a troublemaker and blackballed as a photographer. The local authorities might have even arrested him and dropped him off in the desert just across the California border as so many others had been.

It was not only the industrial accidents that threatened the lives of laborers, but also the generally unhealthy working conditions caused by the disregard for Nevada state mining laws, ones that had long been established to protect, at least in some measure, the health of those working underground. These laws forbade, for example, men working more than eight consecutive hours, or running gasoline engines inside of tunnels without proper ventilation. State and federal regulators, however, never enforced those laws at the dam site, and many men who worked in the excavation tunnels or in the bottom of the canyon suffered carbon monoxide–related health problems. Davis recalled his own experiences being in the tunnels for just a few minutes to take some photographs (figure 5.03). "They worked the men in those

Figure 5.03. "Diversion Tunnels," 1931 (photograph, Winthrop A. Davis). Courtesy of the Boulder City Museum and Historical Association.

tunnels," he said, "and when the oxide gas was heavy, I'd go in and photograph for a half an hour, I'd come out and be so dizzy, I'd have to sit down for a half hour and not know which way to go home."[19]

Labor Strife at Hoover Dam

The excavations of the diversion tunnels (used to reroute the river so that the dam could be built) began on May 14, 1931. The task was incredibly difficult, dirty, and hazardous. Laborers working without hard hats, goggles, facial masks, or any other protective gear had to endure unbearable temperatures and a grindingly arduous task with few breaks and little available drinking water, while continually breathing heavy toxic air caused by inadequate ventilation and an unending line of gasoline-powered trucks that hauled the excavated rock out of the tunnels. Most of these indignities were in violation of mining safety laws, which Six Companies, the State of Nevada, and the Bureau of Reclamation conveniently ignored.

On August 4, 1931, workers were ordered into the tunnels where fumes from a series of blasts to remove rock still choked the air. When they refused, citing safety concerns, they were immediately fired. A few days later, on August 7, 1931, after months precipitated by deaths from explosions, carbon monoxide poisonings, rockslides, and heat prostrations, incidents at the construction site that were compounded by ptomaine poisonings from putrefied meat served in sandwiches, inadequate toilets, bunkhouses with no cooling facilities, and other mistreatments, the men working in the tunnels left their posts and demanded better working conditions, improved living conditions, and "cold water on the job." Six Companies Inc. responded by cutting the wages of all tunnel workers from $5 a day to $4. According to the pro-labor newspaper *Industrial Worker,* which ran the headline "Face Poison Gas or Get Fired," this cut affected not only men working in the tunnels but also "the muckers . . . brakemen, nippers, cable-tenders and cherry picker men."[20]

On August 8, 1931, a general strike was called against Six Companies Inc., and laborers at Hoover Dam walked off the job while adding to their demands a base salary of $5 a day for "surface men," $5.50 for tunnel laborers, and $6.00 for miners and carpenters, in addition to an eight-hour workday that included transportation time from the camp to the job site, a flat rate of $1.50 for meals, enforcement of Arizona and Nevada mining laws, and no discrimination or retribution against strikers. When the strike spread to the entire labor force, Six Companies suspended all operations at the dam "until conditions and the attitude of labor permit us to resume." However, since the project was actually several months ahead of schedule at the time, William H. Wattis, president of Six Companies, Inc., stated, "We can easily afford to wait for better conditions."

The strikers, demanding an investigation by the Bureau of Reclamation into unfair labor practices and hazardous working conditions, arranged for a meeting on August 13, 1931, with Walter R. Young, the bureau's head construction engineer. But fearing the interference of the Industrial Workers of the World (iww) union, otherwise known as the "Wobblies," Young stated that he was not there to investigate labor conditions but only to ensure that work on the dam resumed. The workers themselves did not want to be associated with the iww or its leading organizer, Frank Anderson, who was in Las Vegas to recruit members into the union and agitate support for the laborers' plight but was jailed by local authorities on charges of vagrancy. The laborers working at the dam, however, feared that they would lose their jobs entirely if they were affiliated with the Wobblies, so they held a separate vote specifically to disassociate themselves from the iww.

The laborers meanwhile were not in a strong bargaining position. According to Winthrop Davis, it was the national state of depression and the threat of losing a paying job (however dangerous) that kept many men going back down into the canyon and tunnels day after day. "You know, everybody was worried . . . about the Depression," Davis remarked. "They thought very seriously about their jobs because some of them didn't have a square meal a year before they got onto this job. That's why they stood the abuse that they did. You couldn't work men today like that for $4 a day. You'd have to pay them three times that an hour to go underground and work in those tunnels."[21]

After Walter Young met with the strikers, he ordered Six Companies to resolve the labor dispute and restart work, giving no indulgence to worker demands. Six Companies responded by sending in strikebreakers and police to roust the men out of their bunks in the middle of the night, firing some and detaining others. According to a report in the *Milwaukee Journal*, "The strikers were ousted Saturday at midnight from their living quarters by order of officials of the contractors. The men were warned that any outbreak which would endanger government property would bring federal troops from Fort Douglass, Utah."[22] To illustrate the lengths that Six Companies and local authorities were willing to go to crush striking workers, the report continues that Deputy US Marshal Jake Fulmer confiscated riot guns, automatic machine guns, and tear gas guns given to Six Companies by the Las Vegas police department, apparently for use against strikers, suggesting that the strikers were unlikely to gain any sympathy or protection from local law enforcement if efforts to break the strike turned violent or even deadly.

With little hope left, the striking workers made a desperate appeal to the US secretary of labor, William Doak, asking for his intervention on their behalf, a request that was summarily refused. On August 16, the strikers, acknowledging defeat, voted to call off the strike and return to work. Although Six Companies maintained the pay cut, they did engage in some efforts to improve working conditions and speed the construction of living quarters in Boulder City.[23]

A Model of Decorum

From its inception, the government-built construction camp at Hoover Dam—later named Boulder City—was envisioned by Secretary of the Interior Ray Lyman Wilbur as a place to combat the social ills developing in Las Vegas. Wilbur imagined that Boulder City would be a moral utopia in the desert. A 1930s-era promotional booklet *View Book of Boulder [Hoover] Dam: World's Biggest Job* describes Boulder City as "the most elaborate, modern and comfortable construction camp ever built," while a retrospective on Boulder City in the October 22, 1946, edition of the *Las Vegas Review-Journal* described Wilbur as "a pious man" disturbed by the moral excesses of the roaring boomtown of Las Vegas who "decreed that a new city should be established which should be a model of decorum and physical beauty . . . his dream community."[24]

Established as an official "reservation" by the federal government, Six Companies Inc. controlled Boulder City, which ensured, at least on the surface, Wilbur's desire of an oasis of morality in a sea of Las Vegas sin. Not only was Boulder City a dry town, but being mindfully and proactively anti-union from the start, Six Companies selected Sims Ely to run the city, not only to ensure that the dam workers had a safe and virtuous place to live, but also to guarantee that no union organizing occurred. Ely, described as a "banker-businessman-bureaucrat," managed Boulder City as a benevolent dictator of sorts, acting as a glorified town magistrate who granted divorces, awarded custody of children, jailed "vagrants" and other scofflaws, fixed

Figure 5.04. "Hoover Dam—labor issues; strike breakers," 1934 (photographer unknown). *Left to right:* unidentified; John Andrew Jensen Sr.; unidentified; "Mountain Man" Dean (well-known wrestler, real name Frank S. Leavitt); unidentified. Courtesy of the Boulder City Museum and Historical Association.

prices in the stores, and issued business permits to applicants who met his personal standards of character, personality, age, physical and financial fitness, and business experience.[25] In an attempt to keep the ills of Las Vegas from spreading to Boulder City, Ely instituted strict standards of morality and propriety such as dress codes for citizens and outright bans on alcohol consumption, gambling, and prostitution.

Also illegal were union activities, any suspicion of which would result in termination and expulsion from the city. Ely's "bulldog sheriff" Bud Bodell, who also surreptitiously ran a small gambling racket under Ely's watch (giving Ely and Bodell a monopoly on illegal gaming in Boulder City), was put in charge of enforcing the openly antiunion policies (what some called a "police-state") by firing, forcibly evicting, and exiling anyone suspected of union agitation. Though never officially acknowledged, Bud Bodell went so far as to bring in permanent strikebreakers as "muscle" to be a present reminder to laborers that they should not try to unionize. Unpublished, archival photographs show Frank Crowe and others posing alongside brothers John, Fred, and Ted Jensen, as well as one "Mountain Man" Dean, a well-known professional wrestler (figure 5.04). According to John Andrew Jensen Sr.'s grandson, John "Jack" Jensen Garner, Bodell hired his grandfather and brothers along with wrestlers and others as strikebreakers to enforce his antilabor atmosphere.

A Strike Unseen

Although the strike of 1931 at Hoover Dam has been extensively documented, what is interesting is that while it garnered headlines across the country, there are no published images of the incident in national news or governmental publications. The one or two images that do appear are in small, pro-labor or mining newspapers. Moreover, none of the previous historical accounts of the strike mention this fact.

Winthrop Davis, in keeping with his desire to document the stories of working-class people, is the only photographer I have found to capture images of striking laborers. What is ironic about his images, however, is that instead of exhibiting raucous and tumultuous unionist iconography, they portray the strike as undeniably boring. Instead of ferocious clashes between protesters and police, we see hungry men milling about waiting for a ladle of stew. Instead of masses of people marching and holding picket signs, we see groups of men standing around in the sagebrush, kicking the dirt, or reading the newspaper. Instead of rallies and violent protest, we see men who are simply idling, not *doing* anything. Although we know from various first-person and journalistic accounts that there were certainly clashes, some of them violent, there is no visual record of those events, and the visual record that we have with Davis's photos shows the strike to be mostly tedious.

A testament to the control that the bureau and Six Companies had over the flow of information, the visual record we do have tells us little about the strike and occasionally raises more uncertainties. One example is an image by Winthrop Davis in which he depicts some sort of meeting among strikers, but which also leaves us with many unanswered questions. In this photograph nine individuals are atop a platform in the center of the image (seven standing, one crouching, and one sitting on the edge of the platform, hunched over and resting his arms on his legs), surrounded by a large group of men (figure 5.05). The individuals standing in the center are middle-aged, some graying, most with light-colored pants, and three with ties, one with suspenders and a bowtie—an unlikely appearance of a laborer. Unfortunately, there is no verbal account of what the photo is depicting. Who are these men? Who is addressing the crowd? What is the occasion? What topics are being discussed? These are all questions for which we have no answers.

Curiously, only one man, in the lower center of the image, is turned to engage the camera. Everyone else is either focused on the speaker or, like the man reading the newspaper, seemingly bored. This man appears to be suspicious of the photographer, but for what reason we don't know. Six Companies was known to deploy undercover strikebreakers who infiltrated the labor groups in order to report union activities to Crowe, Young, Ely, and Bodell. Was this man one of them? Conversely, the strikers themselves were distrustful of photographers, perhaps concerned with being documented and identified as strikers and thus facing reprisals later on. Davis commented on those suspicions, saying of his strike photographs, "If I hadn't been in good relations with the working men, I wouldn't have gotten the pictures that I got. They were just about ready to mob me as soon as I set my camera up. Then some of the boys would come in and say, leave him alone,

Figure 5.05. "Labor Strike, Aug 7–13," 1931 (photograph, Winthrop A. Davis). Courtesy of the Boulder City Museum and Historical Association.

he's with us. I told them that you guys aren't going to get anywhere unless you get some publicity on this job."[26]

Not surprisingly, Davis's assessment that the lack of photographic coverage likely contributed to the low level of public support for the strikers was prescient. We might never know the answer to who was speaking on the platform, or the identity of the mysterious man looking at the camera, but one wonders whether things might have turned out differently for the laborers had more strike-related photographs made it into circulation. It would be speculative to attribute the tepid public support for Hoover Dam laborers in their struggle for safer working conditions to the lack of photographic coverage—or more specifically the lack of *dramatic* photographic coverage—but considering the social impact of images such as Milton Brooks's 1942 Pulitzer Prize–winning photo "Ford Strikers Riot" published in the *Detroit Free News*, showing a cowering union member being viciously clubbed, punched, and kicked by six gleeful strikebreakers, one might wonder what would have happened had similarly unsettling images of the Hoover Dam strike been splashed across the pages of national newspapers.[27]

Women Seen and Unseen

Though women, too, are largely absent from the visual record, the March 31, 1928, edition of the *Los Angeles Evening Herald* features one of the most spectacular depictions of the project (figure 5.06).[28] Purchased by William Randolph Hearst in 1922, the *Herald* appealed to blue-collar readers by offering a steady stream of sex, crime, and celebrity scandals.[29] The chief competitor of the *Evening Herald* was the *Los Angeles Times*, a morning newspaper that promoted Los Angeles and Southern California as a paradise destination for anyone interested in leaving the cold, snowy climates of places like Detroit, New York, and Chicago. The publishers of the *Times* and *Herald* (and, to varying degrees, the four other Los Angeles daily newspapers in the 1920s and 1930s) were fully aware that the development of the region and the profitability of newspapers were mutually dependent, and so they actively promoted Southern California as nothing short of "Elysium."[30]

The boldfaced and centered edition title of the *Evening Herald* marks its intentions unambiguously, but the illustration's mix of art nouveau and modernesque design styles suggests a more subtle and symbolic argument as well.[31] Both the woman and the page's border adornments are highly reminiscent of art nouveau (which was several years out of fashion by 1928),

Figure 5.06. "Boulder Dam Prosperity Edition," *Los Angeles Evening Herald,* March 31, 1928. Courtesy of the Boulder City Museum and Historical Association.

while the dam and skyline are art deco.[32] Because of its mixture of lavish nouveau symbolism and images of modern transportation and industry, the whole composition suggests a transition from the antiquated assumptions of the late nineteenth-century art nouveau era to the machine-age modernism of the early twentieth century. Moreover, the image is remarkable for its encapsulation of several major themes of Hoover Dam imagery: economic prosperity, the value of nature, modern technology and progress, and the representation of the dam's massive scale. While the visual rhetoric surrounding the dam clearly participates in well-established and characteristically American traditions involving the technological domination of nature, the *Evening Herald* image straddles major artistic vernaculars as the relationship between nature and technology is presented in ways that both echo and challenge long-standing American conventions.

The transitional style of the illustration can also be viewed as emblematic of the complex relationship between nature and culture, and between the natural environment and the man-made environment. Organic forms such as vines, flowers, and leaves dominated everything in art nouveau style from jewelry to glassware to idealized images of women, and we see this displayed in the depiction of the woman, as well as in the fruits, flowers, and leaves scattered throughout the image.[33] As James Grady, the first person to catalogue the importance of nature in the aesthetic expression of art nouveau, argues, the principles of nouveau expression were defined according to human needs but were also found in nature; the reversion to nature is what inspired the best nouveau art and architecture of the period and is what gives it its distinctive style, as nature's rhythms were thought to have an endless capacity for illustration and dramatic power.[34]

The composition of the page draws attention to the dam by placing it in the center of the page, but the true focal point of the illustration is the colossus-like woman standing at the edge of the water, her robes swirling into an overflowing cornucopia as she releases handfuls of fruit that descend into the river and fade into imagery of technology, architecture, agriculture, and commerce.[35] This image depicts a woman who embodies a complex nexus of contradictions between modernist and traditionalist ideals and artistic conventions. The woman is a figure in harmony with her surroundings: evocative of nature, symmetrical, and the subject of intricate composition, yet also comfortable being completely enveloped by technology. Her gaze shows an inner confidence, directness, and a sense of self, deriving her power from nature and the symbols that surround her. Although the woman seems to be embedded in place like an artful statue perched precariously at the edge of the water, this type of positioning would not be unusual for art nouveau artists, as they often depicted highly idealized women and androgynous males as similarly postured in precarious positions. As such they make for attractive decorations without threatening anyone's manliness or heterosexuality.[36]

But not everything about the woman adheres to archetypal nouveau design. Art nouveau is often identifiable by its depiction of provocatively dressed (or half-dressed) women placed in alluring poses. Consequently, one might, for example, expect the woman to be laying seductively on her side across the top of the dam while dropping her fruits into the river. Instead, she is depicted standing, dominating the image. And although her commanding size casts the men and their technology as subordinate, this is not altogether unique in nouveau design either. For instance, Henry Somme's nouveau-style art of the late 1800s includes a series of drawings in which Amazon-sized women handle men like marionettes. Other illustrations published in popular print at the time show miniature men held captive or performing tricks for giant Amazons, or depict the woman as domineering, which is described by Elizabeth Menon as a "prototype of the new 'modern woman' as envisioned by the male illustrators of the popular press."[37]

But contrary to nouveau artists' tendency to portray women as submissive objects of desire, this depiction suggests certain progressive ideals that allegorize representations of women standing or striding in flowing robes to signify modernity, progress, or justice.[38] If the woman's exaggerated size symbolizes a new "modern woman," the loose-fitting robes also suggest a woman liberated from the constricting clothing styles of the late nineteenth century and convey a quality of free-form movement reminiscent of Loie Fuller.[39] Fuller also transcended the larger cultural movements of the late nineteenth and early twentieth centuries by integrating the female body and abstract natural metaphors of art nouveau with machinery, advanced lighting and stage design, and an embrace of modernism and technology.[40] Fuller, an internationally popular dancer known for mesmerizing performances in which she would dance in hundreds of yards of silk and cloth illuminated by a complex array of electric colored lights, provided an image of a new, modern woman who moved beyond the art nouveau ideal to become signifier of modernity and national progress.[41]

Another chief feature of the woman is her cornucopia. Notice, however, that she is not holding a cornucopia; instead, her robe coils into a cornucopia-like shape filled with fruits, suggesting that the woman, an allegory of the life-giving river or nature itself, has the capability to bear fruit— technology, industry, and commerce—thus bringing the machine age into existence. The cornucopia, also a symbol used throughout the late 1800s and early 1900s to promote the State of California, brings to mind the Roman goddesses Aequitas, a goddess of fair trade and merchants, and Abundantia, a goddess of good fortune, abundance, and prosperity, who are typically depicted with a cornucopia.[42]

Art nouveau as a movement was finished by roughly 1910, and it was then that machines replaced nature as the new artistic and architectural inspirations for the Bauhaus, modernist, and International styles that took hold in the 1920s.[43] The mixture of design influences in the *Evening Herald* illustration suggests a cultural transformation from one in which nature was the predominant inspiration to one that was invested in new machines and technologies, as well as in new forms of art and architectural expression.

Art deco designers in the 1920s attempted to capture the spirit of the machine age through styles like Streamline Moderne, for example, which offered a near-utopian, sci-fi, fully automated world in which machines, controlled by man, were ubiquitous. Sleek curves, teardrop shapes, small rows of windows, straight horizontal lines, and shiny metallic membranes that conceal the inner workings of the machinery typically identify Streamline design. The ultramodern aesthetic of Streamline displayed an intense fascination with speed, and its visual vocabulary consisted of a number of motifs drawn from forms associated with high-speed modern transportation.

At the bottom of the image we can see the fruits of the woman's (nature's) bounty transformed into machines of many types: boats, ocean liners, tractors, trains, industrial buildings, and skyscrapers. Although the machine imagery is not specifically Streamline or futurist, it does convey the same principles in depicting fast, powerful, modern means of transportation. The train, for example, is drawn with striated horizontal lines to suggest that it is moving swiftly, and the large ocean liner, the quintessential emblem of luxurious transportation in the 1920s, is shown from the front, highlighting its sinewy contours instead of its broad profile.

Allegory and Gendering in the "Evening Herald" Image

Robert Hariman writes that allegories arise when two seemingly incommensurable realms collide, creating a space for the juxtaposition of ideas that must somehow coexist to be able to make meaning within widely known conventions.[44] This conception seems particularly applicable to the illustration at hand. Like words and images, allegories are the result of dialectical cultural experiences that must occupy the same space but yet are somehow incommensurabilites that occur at the confluence of cultural transformations in which, with the Hoover Dam, for example, the technologies of significance begin to overwhelm even the most sublime natural wonders. Allegorical manifestation as visual personification is one way that we come to terms with these types of monumental shifts in science, culture, society, and intellect.

Like the *Evening Herald* image, an editorial cartoon published in the September 30, 1935, edition of the *Los Angeles Times* titled "The Gateway of Empire!" echoes these themes of allegory (figure 5.07). In the image, a female personification of the United States raises the flag of Hoover Dam—a symbol of a new nation-state or new sovereign territory that the dam is creating—while her sash with the words "The Great Southwest" billows in the breeze. The woman is clearly reminiscent of the Statue of Liberty or

Figure 5.07. "The Gateway of Empire!," *Los Angeles Times,* September 30, 1935 (illustration by Bruce Russell). Copyright © 1935. *Los Angeles Times.* Reprinted with Permission.

Lady Liberty as suggested by her radiate crown, but owing to the design of her chemise, she also bears some resemblance to images of *Libertas*, the Roman goddess of freedom. Comparable to other mythic female personifications of America, and similar to the woman from the front page of the *Evening Herald*, this female character is a triumphant, nationalistic emblem. While the Statue of Liberty symbolizes the gateway to the United States for millions of European immigrants, the illustration shows Hoover Dam as the new gateway for expansion into the Southwest. Reflecting previous associations between the dam project and empire, both the rendering of the flag and the use of the actual term suggest that the dam's construction signals the expansion of American imperial sovereignty over what was once desert but will soon be transformed into fertile farmland and home to millions of new residents.

The history of this kind of personification dates to classical antiquity, in which abstract concepts such as wisdom, abundance, art, the seasons, geographical regions, and nature are represented as living females.[45] For example, the nine Muses of Greek folklore that inspired artists and intellectuals were, of course, all female. Likewise, women are almost always depicted as the physical manifestation of political ideas such as freedom, republic, or democracy. Allegorized women have also represented "tradition" and "progress," two somewhat incongruous terms that have, in reality, been the purview of women and men separately, with the woman practicing tradition and the man practicing progress. When new ideas have come into existence and the choice is made to depict that idea in personified form, it has invariably been as a woman. Moreover, we can see in these allegories patrimonial tendencies of disciplining the spectrum of female personifications of human culture.

The irony here is that although personifications of women represent notions of art, politics, and science, in reality women have historically been excluded from practicing those same vocations. For example, Hygeia, goddess of health and daughter of the god of medicine, Aceso, the goddess of healing, and Panacea, the goddess of remedy, are all female personifications, yet females were barred for many years from pursuing medical training outside of midwifery. The woman in the *Evening Herald* image towers over the accoutrements of power, science, commerce, and bounty; however, throughout history women have rarely been able to wield the scepters of power, engage pursuits of science, or stand with the titans of industry.

Even seemingly mundane associations with cultural norms and expectations such as work attire become totalizing iconographic symbols of the relationship between the state and women. In allegorized presentations such as this one, for example, the woman is invariably shrouded in robes but never practical or professional clothing. Whereas representations of soldiers might depict them in their field dress, or tributes to labor as workmen with their boots and hardhats, women almost never wear professional clothing and instead are portrayed barefoot with flowing robes without any items of dress or equipment typical of their vocation.

By being presented as allegory, we also see women as powerful, but not authoritative. In nearly every society, men are at the center of cultural value and authority and have some culturally sanctioned authority over women. Although women are far from powerless, they have a clearly defined role of subordination and compliance and typically have little authority over men, the distinction here being that power and culturally legitimated authority are not the same thing. Max Weber proposed the classic distinction between power and authority in that authority is a right to make decisions and demand compliance, whereas power is the ability to effectively influence persons or things that are not necessarily allocated to those individuals or their roles. Authority is hierarchical, while power is exerted through influence outside of chains of command and control. In other words, power (in its various guises) is the ability to gain compliance, whereas authority is the social recognition of those in influential positions.[46]

But allegorical gendering is not limited to just notions of science, politics, or industry. As Carolyn Merchant and an entire school of ecofeminist thought remind us, gender is encoded into the narratives and metaphors of landscape as well, and is a crucial consideration for any understanding of Western depictions and narratives of nature.[47] Here we see the gendering of nature as female. In Western culture, feminine nature typically appears either as a virgin, showcasing her potential for development, or as wild, thus requiring reclamation or improvement.[48] There is, however, a third feminine nature, "Mother Earth." The notion of earth as a "Mother" has been with us since at least as far back as the ancient Greeks, with Gaia bringing forth the earth from the void, but in the *Evening Herald* image we see Mother's nurturing garden improved by technology showering her bounty into the river and onto the men and machinery at the bottom of the frame. In this image we see nature as both providing and passive. This stands in stark contrast to the way that nature—the Colorado River, for example—is portrayed in nearly every account of the project. Descriptions of the river range from untamed, to threatening, to downright menacing, while the surrounding desert is merely, as President Franklin D. Roosevelt described, a "cactus-covered waste."

The *Evening Herald* image also recalls biblical narratives. The woman evokes the story of Eve as she showers fruit upon the smaller, unsuspecting male figures at the bottom of the frame. Technology, and man's use of technology to dominate nature, has also been, and continues to be, sanctioned and inspired by the Genesis creation story, according to which man was able to lounge about in Eden without any need to disturb the natural landscape because everything—food, clothing, and shelter—was provided. In the image, the farmer on the far left of the page is not shown with ox and plow, but instead is riding a tractor: an iconic symbol of modern rural life and American agrarianism. This parallels the Protestant notion that once cast out of Eden man became dependent on the technologies needed to cultivate the natural landscape for his survival. In this way, technology is seen as a divinely instituted ability and right of man in his constant quest to re-create the Edenic state from which he has been expelled.[49] The analogue here is that patriarchy invokes a notion of conquering, mastering, manipulating, and exploiting what men perceive as less powerful than themselves.[50]

Acknowledging the gendering of nature in this image can also reveal certain characteristics that are hidden or concealed in the presentation of the project. One of the central characteristics of this image is its near singularity. To my knowledge, this is one of only two pre-construction images in which a female and the project are depicted together, and the *only* image that features both a representation of a female and the dam itself. In the case of the Hoover Dam project, females are conspicuous by their absence, which only serves to enhance the sense of patriarchy that permeates nearly all Hoover Dam imagery up to roughly the early 1940s. In fact, Hoover Dam imagery rarely depicts women engaged with technology of any kind, a circumstance that speaks to the overarching tone of the dam's imagery taken as a whole: as women are conspicuously absent from nearly all visualizations of the dam, their scarcity serves to further reinforce the hypermasculine portrayal of the project.

TOP: *Figure 5.08.* "Mother and Three Children in Las Vegas Hooverville," ca. 1931 (photographer unknown). Courtesy of the Boulder City Museum and Historical Association.

BOTTOM: *Figure 5.09.* "Prostitutes at the Arizona Club on Las Vegas's Block 16," ca. 1931 (photograph, Winthrop A. Davis). Courtesy of the Boulder City Museum and Historical Association.

Visual and Verbal Depictions of Women

Although women were largely excluded from images of the dam itself or technology of any kind, those who lived at or near the construction site were occasionally depicted in other images. Nonetheless, these were often restricted to a binary of patrimonial depictions that mostly frame the woman as a personification of the cultural ideal as a mother or caretaker, but occasionally as the *fille de joie*. This dichotomy is one that conveniently fulfills and reifies the dominant patriarchal affirmation of woman's intrinsic duality: on the one hand a matron suited for child rearing, and on the other a woman suited for pleasure.

Photographs of women as matrons almost always show them at home (often just a tent) tending children. In figure 5.08 we see an image representative of the typical depiction of women and similar in many ways to the "Williamsville" image discussed later. Here we see a mother who lives in a tent holding one child while her other two children play in the dirt. A small coal stove sits next to the tent, and a table and various belongings are strewn about behind them, while in the distance is a white home. Just as in the "Williamsville" image, there are no men present in the photo, as men are almost never shown in the same frame as children. Likewise, there is also no machinery or dam-related technology in the photo, as is also characteristic. Other images depict women in front of temporary housing, children playing in the scrub grass and desert sand under a woman's supervision, and women cooking, cleaning, or tending children in a variety of similar settings.

Another image of women is one of Winthrop Davis's most well known photographs, "Prostitutes at the Arizona Club on Las Vegas's Block 16" (figure 5.09), which generated so much interest because it expressly violated the publically acceptable narrative of woman as matron and instead flaunted taboo through its (scandalously for its time) direct and unapologetic depiction of prostitutes. In this image we see Las Vegas's designated "red light" district, called Block 16, where brothels, bars, speakeasies, and gambling halls would entertain many of the thousands of men who worked at the dam site. Prostitution was rampant in Las Vegas but banned in Boulder City; however, Davis describes how it moved with ease from one location to the other. "When the construction started the government started squeezing the guys in Vegas to hold it down," he said. "Well, half of those prostitutes left Vegas and they came out here and they were supposed to be wives of some of the workers. That caused quite a bit of trouble for awhile cause some of the women were working two shifts, if you know what I mean."[51]

In keeping with the government's campaign to curtail visual and verbal testimonials that ran counter to their preferred narrative, Davis's now widely seen photograph of prostitutes did not find circulation at the time of its production. Only later in efforts to showcase his work in galleries and art museums did these photographs receive public airing.

But while women were often excluded from photographic depiction by (male) photographers, a few women did seize their own agency and wrote about their experience at the dam. In a 1931 article titled "What Reclamation Women Do in Las Vegas," D. L. Carmody, wife of engineer Don Carmody, echoes many of the prevalent depictions of women as subservient to men and struggling for an independent identity. Reverting to the cultural ideal—mother—she writes with religious-like fervency that the role of an engineer's wife is to follow her husband as he goes about "bringing order out of chaos, turning life-giving waters on arid lands, opening new fields for homeseekers and settlers and by the might of his shining vision bringing beauty out of desolation." She describes an unglamorous life of moving year after year, planting quick-growing vines, and then packing and unpacking again. To her, the role of the woman was to be "domestic" and a "home-loving body whose affections cling about the family hearth-stone." She describes the wife of an engineer as a woman whose "mission in life consists of following him to the far corners of the earth." But the Hoover Dam project was an opportunity to live in one place for four or five years, a welcome luxury for a family tethered to a career that requires constantly moving to the next project locale. Yet, she describes the social scene among wives of reclamation employees who accompanied their husbands in the move to Las Vegas as springing seemingly from sheer boredom.[52]

In another article written by Carmody titled "Woman's Impression of Boulder Project Told By Wife of Engineer" (a retitled reprint of the same article published in the journal *New Reclamation Era* the previous month called, simply, "A Visit to the Hoover Dam"), the article's title caveat reminds us that it is told not just by a woman, but by the wife of a (male) engineer. And while there is nothing in her account that suggests a particularly "womanly" point of view, her descriptions are vivid and at times echo others that portray the project as a sublime undertaking. Carmody provides a colorful glimpse into what the activity was like at the dam site in 1931. According to her account, the apt descriptor of the project is "activity" as she describes the continuous motion of "men climbing perpendicular ladders five or six hundred feet to the top; men clustered about the mouths of embryo tunnels; men swinging across the turgid stream in little cars attached to strong cables; men and machinery going into places where even a mountain goat would hesitate to climb." But one passage differentiates her perspective from many others of the time. At the end of the article Carmody wonders what this project will mean for generations to come. Unlike Oskar J. W. Hansen, who wrote that the dam would be "imperishable against the oblivion imposed by time," Carmody wonders what will remain of all the work and struggle the men are enduring to build this monument to modernity. She writes,

> And we know that in the end man will conquer; but by some alchemy hard to understand the scene remains unchanged and this advance of men and machinery seems but surface activities like the clouds passing over a landscape. . . . Some day the Hoover Dam will be completed, an accomplishment of such magnitude as to place it among the wonders of the world . . . but in the years to come, the countless thousands of years which may lie ahead of us, what of this man's work will remain? What will another generation, a strange race of men, glean from the mighty dam stretching across the stream? What will these mountains tell the unknown future of the forgotten past? I wonder!

African Americans

When a massive public works project to build a dam on the Colorado River outside Las Vegas was announced, it promised to bring with it an unprecedented employment boom to Las Vegas during the worst economic climate in US history. When actual construction started on Hoover Dam in 1931, there were approximately seven million unemployed Americans, roughly forty-two thousand of which found work on the dam at one time or another. But due to discriminatory hiring practices and worsening race relations in Las Vegas throughout the early 1900s, few blacks were among those who found work at the dam.[53] In fact, the *Las Vegas Age* wrote in February of 1932 that once the dam was completed "an average number of 4,000 employees will have rolled up the stupendous number of 71,500,000 man-days." The average description of such employees, the article continues, will have been "37 years of age, white, [and] American born." Unfortunately for blacks seeking work, very few of those man-days were available to them.[54]

Blacks have lived in southern Nevada since the late 1890s, when they helped build the railroad into the dusty outpost that was later to become the city of Las Vegas, which was incorporated in 1905. But from its very beginnings, the black community encountered segregation, racism, and discriminatory hiring practices that would prove devastating to their quest for social and economic equality. Although there were few blacks or other minorities living in the area at the time, the Las Vegas Land and Water Company, a subsidiary of Union Pacific Railroad that owned the land parcels that were auctioned off during the city's enfranchisement, felt it important to take precautionary measures by proactively segregating blacks and other minorities to Block 17 (adjacent to the Block 16 "red light" district) in order to prevent them from becoming scattered throughout the city.[55] Moreover, blacks were disqualified from purchasing property or opening business in any other section of the city for several years thereafter.

One extreme example of this policy was attempted by Utah land investor Tom Williams, who, in an effort to establish his own entirely white town, purchased 150 acres for a new town site in the area that would eventually become North Las Vegas and proudly announced it to be a place of "no taxes, no licenses, and no Negros."[56] Blacks were barred from purchasing property there, too, a practice that would continue for another thirty years. Ironically, it was not until the move toward total segregation was in full swing in the early 1930s that black businesses began appearing in the "colored section" of Las Vegas as a result of a need for services by blacks and to appease white officials by keeping all of the black-owned businesses and their clients in one area.[57]

Although life was difficult for blacks in Las Vegas, they continued living in the area even as discrimination was mounting throughout the early 1900s. The Ku Klux Klan held a march down Freemont Street in 1924, and although the Klan did not make the inroads it wanted in Nevada, likely because of the sparse population of blacks throughout the state, it did bring the specter of racism to the forefront in the city. The continued influx of migrants from southern states also added to the racial oppression of Las Vegas, as they brought with them the segregationist and bigoted attitudes of the Deep South, views that went unchallenged when they arrived in Las Vegas.[58]

When the dam project was announced, blacks, like so many others, moved to Las Vegas with the hope of finding work. But upon arrival, they found the area unwelcoming, as segregation required that they live only in designated areas of the city. Worse, ambiguous and selective "vagrancy" laws formed a system of peonage in which blacks were arrested on trumped-up charges and forced to sweep streets or dig sewer ditches, or were placed on chain gangs. Often the bonds were set at upwards of $250 or more, a sum that violators would never have been able to pay considering that those who did have jobs were typically paid $4 to $5 per day in wages.

Unions had successfully established a presence in Las Vegas years prior to the dam project, and as a result, unionized labor performed most of the preliminary work funded by the Department of the Interior for railroads, roads, bridges, and construction work in Boulder City. However, blacks had been excluded from joining unions since the 1890s and thus were not able to work on any of the preliminary projects, either. In May of 1931, blacks responded by forming their own union, the Colored Citizens Labor and Protective Association of Las Vegas (CCLPA), a group that had 247 members by September of that year.[59]

Although the *Las Vegas Age* called into question the legality of antiblack hiring practices by reporting that various construction programs throughout Las Vegas employed foreign workers but "no Negro labor whatsoever,"[60] Six Companies itself was under no legal obligations regarding hiring other than the explicit language in the government contract that forbade "Chinamen" and required the hiring of US citizens while giving preferences to military veterans. Unfortunately for blacks seeking work, this provided too much leeway in the interpretation of the contracts. For example, although there were many black military veterans who applied for jobs, the preferences for veterans also seemingly only applied to white applicants. One member of the CCLPA wrote to the *Las Vegas Age* in January of 1932 that "there have been since the creation of this association many, many colored overseas soldiers and citizens who have applied in person, with their discharge papers, for work on the Hoover Dam project," but that every single applicant had been denied.[61] And while Six Companies was required by law to give preferences to military veterans, they were also required to hire US citizens, which enabled Six Companies to institute "citizenry tests" similar to the literacy tests, poll taxes, and grandfather clauses that had been responsible for the widespread disenfranchisement of blacks from Reconstruction in the 1870s all the way through the 1960s.

To further insulate themselves from hiring blacks, Six Companies also established requirements for "work experience" that applied only to minorities. The best example of the speciousness of the "experience" requirement was in the hiring of laborers to work in the diversion tunnels. Before construction of the dam could begin, diversion tunnels needed to be drilled through the canyon walls to reroute the river. The work involved some blasting and mining, specific skills indeed, but for the most part the work was simple excavation performed by unskilled laborers whose only job was to work a pneumatic drill and to shovel rock and debris. The excavation of the tunnels alone employed over thirteen hundred men by the end of 1931; however, not a single one was black. In fact, the vast majority of those who worked on the dam had no prior experience blasting, mining, drilling, or anything else related to the project; they simply needed a job. Thus, it was obvious that the question of whether the applicant was a minority or not far outweighed other considerations such as experience. Leonard Blood, deputy director of the Nevada Office of Labor at the time, suggested that one reason for this was that hiring blacks would cause tensions with white workers and that because of existing and accepted segregation practices there would be difficulties and added costs in housing, transporting, and feeding blacks in separate facilities.[62] In fact, a memo to Commissioner Mead from John Page, acting construction engineer, stated that "while Negros would probably be desirable [workers] on account of the extreme heat," the issue of segregated housing has "rendered it impractical to plan on their employment." Commissioner Mead agreed and responded that hiring practices were strictly the jurisdiction of Six Companies and that the construction of separate facilities for persons of color was "hardly feasible."[63]

In May of 1932, the National Association for the Advancement of Colored People (NAACP), after investigating Six Companies' hiring practices, held a meeting with influential members of the Las Vegas community to remind them that blacks were American citizens, too, and demanded that their rights be upheld. The meeting did nothing to change the position of Six Companies, however, since they continued using the "experience" test as a way to avoid hiring blacks. In June of the same year, members of the NAACP and the CCLPA, Senator Tasker Oddie, and Senator Wilbur again met with Six Companies and this time persuaded them to agree to "no further" discrimination, which as a small victory in itself was a tacit admonition by Six Companies that discrimination had, in fact, been occurring. That same month—a year and a half after construction started—the first ten blacks were hired, and by September there were twenty-five. Nonetheless, those numbers were minuscule compared to the workforce of forty-five hundred strong. Even at the peak of employment in July 1934, when, according to the employment office, 5,251 people were working on the dam, there were still only eleven blacks, which amounted to 0.002 percent of the total workforce. Moreover, the few blacks who were hired to work on the dam were relegated to the quarry, one of the hottest and most inhospitable areas of the entire jobsite.

Given this history of discrimination at the dam, one might wonder whether there is any photographic documentation by the Bureau of Reclamation that might provide some insight into the issue of minority hiring practices at the dam. The answer is that the photo in figure 5.10 is the *only* image that shows black workers at the dam. It is but a single—and staged— photograph of six black laborers, one that was never published but kept in reserve to distribute as part of a publicity buffer in the event that charges of racism were brought against the government. Although relatively well known now, there is no evidence that this image ever appeared in dam-related news or promotional publications during the dam's construction.

Figure 5.10. "Negros employed as drillers on the construction of Hoover Dam," October 3, 1932 (photograph, Ben Glaha, Bureau of Reclamation). Courtesy of the US National Archives and Records Administration.

Taken by Ben Glaha, a photographer working under the direction of the Bureau of Reclamation who is responsible for many of the most recognizable images of the dam, the photo shows six black laborers holding various pieces of machinery, while apparently working at the dam.[64] I say "apparently" because the image is curious for several reasons. First, Glaha is best known for his "machine aesthetic" photography of the dam: showcasing its technology, its engineering, its modernity, its sublimity. He took relatively few images of laborers when examined in the context of his entire body of work, although the few he did take have largely become iconographic images in their own right. Second, Glaha and other photographers such as Winthrop Davis did not, as typical practice, arrange or pose men to take photographs. As Barbara Vilander has pointed out, nearly all of Glaha's images of workers are documentary-style photographs taken of workers as they perform their tasks, often unaware of or ignoring the camera. Of Davis's photographs that we have seen thus far, none appear to be posed images.[65] But here we see six men assembled across the frame in a perfect line. These men were clearly stopped from their jobs and brought to this spot specifically for the purpose of being photographed. Third, unlike most images of workers, these men look directly into the camera, and as opposed to other images of laborers, some of these men are smiling, a rare occurrence indeed and one potentially directed by the photographer to show that the project employed not only Negros, but happy Negros. The fact that these men are all directly facing the camera also serves to unambiguously and unquestionably emphasize their ethnicity. Finally, unlike many of the white laborers who worked shirtless to battle the heat and who were photographed as such, none of these men are shirtless. In fact, they are completely clothed head to toe except for bare forearms despite the fact that they are working in one of the hottest areas of the jobsite, which also happens to be one of the hottest places on earth.

It is unknown whether these men were actually employees of Six Companies, but since this photograph was taken in early October, these men could have been employees (though the men assembled and photographed here would constitute 25 percent of the entire black workforce). To highlight the uniqueness of such an assemblage of minority workers, when specifically asked whether he had seen any blacks working at the dam, Winthrop Davis, who documented the project extensively for several years, said, "There were very few colored people, even in Vegas, a few of them in North Las Vegas, but

I never saw any blacks [working] on the dam. I might have seen some but I don't recollect them."

Native Americans

Although proximate human communities such as the Hopi, Paiute, and Papago had lived for generations in the Colorado River basin and the (newly named) Imperial Valley area, they were not included or considered in the Colorado River Compact negotiations.[66] With regard to the Hoover Dam project and the Colorado River Compact, the Bureau of Indian Affairs—the official (and only) governmental body capable of ensuring Native American participation in deliberations—was flagrantly negligent in their duty to look after the interests of native peoples. The lack of a legitimate voice, and thus an inability to contest new laws and land rights agreements, allowed the US government to freely redistribute land to white farmers without opposition. For example, thirty-nine thousand square miles of land were set aside as a Moapa Paiute tribal reservation by the federal government in 1873, yet just two years later, the reservation was reduced to a mere one thousand acres. To put that in perspective, that is a reduction of nearly twenty-five million acres, or 99.9996 percent.

The justification for terminating native land rights was that the reservations were supposedly trapping native peoples in a cycle of poverty; therefore, it would be better for them if they were simply "assimilated" into mainstream American culture. However, the assimilation and redistribution plan was not exactly equitable and instead heavily favored white farmers over native farmers. Donald Worster writes of an example on the Yuma reservation in which "the native owners were allotted a mere 5 acres apiece . . . while the 6,500 acres left over from the distribution were sold to whites in 40 to 100 acre parcels."[67] This not only served to limit the land and political and economic influence available to natives but also reduced native water demand and possible claims to rights of prior appropriation whereby native peoples could declare "beneficial use" of the water and thus reduce the amount available for white farmers.

Historically in the United States, the treatment of Native Americans has many parallels to that of African Americans, and, interestingly, their depictions at Hoover Dam (at least in photographs) closely mirror each other as well. While a large population of Native Americans lived in the region, only a few were hired to work on the dam project, and all were given the dangerous job of high-scaling, supposedly because of the myth that they were not afraid of heights, a notion that has its genesis in the building of the Canadian Pacific Railway bridge from the Kahnawake Mohawk Territory across the St. Lawrence River to Montreal Island in 1886. On that project, Mohawk from the Kahnawake worked hundreds of feet off the ground during the bridge's construction, and then continued to do similar work on skyscrapers in New York and other places, thus establishing a reputation that became generalized to all Native Americans and ingrained as part of their cultural mystique among whites.

Although only one photograph exists of black workers at the dam, there are a small number of photographs of Native Americans.[68] Two of them are staged almost identically to the black workers' image, and if government records are accurate, they were taken only two days after the black workers' photo.

The photos shown in figures 5.11 and 5.12 were taken on October 5, 1932. The fact that they were taken on the same day, yet in somewhat different

Figure 5.11. "Indians employed on the construction of Hoover Dam as high-scalers. This group includes one Yaqui, one Crow, one Navajo and six Apaches," October 5, 1932 (photograph, Department of the Interior, Bureau of Indian Affairs). Courtesy of the US National Archives and Records Administration.

configurations, suggests a similar motivation to that of the back work-ers' photograph. The rationale for these particular images is unknown, but since images of laborers aligned in a row for the purposes of a photograph are nearly nonexistent, one might surmise that these were a similar hedge against claims of racism or inequitable hiring practices.[69] Like the image of the black workers, none of these images appear in a publication from the first half of the twentieth century, to my knowledge.

In figure 5.11 we see nine men depicted in such a way as to play on the notion of their inherent comfort with heights, as they are seemingly poised on the edge of a cliff, with the void of the canyon as the backdrop. This image also alludes to their job, the dangerous high-scaler who rappelled down the sides of the canyon to place blasting dynamite in the walls and to loosen rock with pry bars. Like in the black workers' image, these men are arranged in a group looking directly at the camera. All of the men are wearing hats, with three of them shown in "hard hats," or at least what passed for head protec-tion for those working on the dam.

In figure 5.12 we see two of the men from figure 5.11 (the man kneeling on the far left in figure 5.11 is now without his hat and is standing and point-ing, and the man kneeling on the far right in figure 5.11 is lower in the frame in figure 5.12, still wearing the same hat). This image is one that plays on Western iconography of Native Americans depicted as guides for European explorers. These kinds of images appear in such works as Maurice Thomp-son's *Stories of Indiana* (1898), in which native guides are sketched allegori-cally pointing westward for Robert de La Salle's exploration of the Great Lakes region of the United States and Canada, or the statue of Sacajawea by sculptor Alice Cooper unveiled in 1905 at the Lewis and Clark Centen-nial Exposition. The bronze cast sculpture, which now resides in Washing-ton Park in Portland, Oregon, depicts Sacajawea with her arm raised look-ing high off into the distance, symbolizing her role in the exploration of the North American continent and the subsequent great westward migration by European settlers. This image of the two native Hoover Dam workers is clearly staged to draw on that iconography and to symbolize the native people's continued role in aiding the expansion of white dominance in the American West.

The next photograph reproduced here is of Moapa Paiute from a series of images taken in 1925 (figure 5.13). Although Paiute had been photographed as far back as the 1870s, when Timothy H. O'Sullivan famously documented the

Figure 5.12. "Apache Indians employed as high-scalers on the construction of Hoover Dam," October 5, 1932 (photograph, Department of the Interior, Bureau of Indian Affairs). Courtesy of the US National Archives and Records Administration.

American West as part of Lt. George M. Wheeler's survey of the continent west of the One Hundredth Meridian, this is one of the earliest known photographs of Native Americans shown in connection with the dam project.[70] The photograph in figure 5.13 is one of several images taken on the same day in 1925 showing Nevada governor James Scrugham with Moapa Paiute Indians scouting Boulder Canyon. Here we see Governor Scrugham standing and looking face-to-face with one of the Paiute. Aside from the oddly confrontational stance of the governor and the fact that he is not looking at the camera but instead stares intently at the Native American beside him, the difference in their attire is probably the most striking feature of the image. The caption states that these are Moapa Paiute Indians "in costume." Although the purpose of these images is unclear, based on other photographs of the time, these Paiute seem to be dressed in attire atypical of their time. Other contemporaneous images as far back as the 1870s show Paiute (and other tribes) dressed in something of a hybrid style of settler/explorer and native attire. In O'Sullivan's well-known photograph "Pah-Ute (Paiute) Indian group, near Cedar, Utah, in 1872," for example, we see Paiute dressed in pants, hats, button-down shirts, long overcoats, and scarves. In other photographs in the early 1900s we see similar styles of dress. For Paiute to be depicted as they are shown here, yet more than fifty years after O'Sullivan's photo, leads one to conclude that the image is meant to visually communicate that those shown with the governor were, in fact, "real" Native Americans. Like other images of Native Americans, I have not found any publication in which these images appeared, and so their actual purpose remains unclear.

Like the images of black workers (and all other images of laborers), and unlike nearly all of the politicians, engineers, and government and Six

Figure 5.13. "Boulder Canyon—Nevada Governor James Scrugham with Moapa Paiute Indians in costume at the USBR gauging station," 1925 (photograph, US National Park Service). Courtesy of the Boulder City Museum and Historical Association.

Companies officials, none of the Native Americans in these photographs are named, and they are only occasionally labeled by their tribal affiliation. Furthermore, although I discuss this at more length in the next chapter, it is important to note here that these types of images of laborers in which they were photographed up close and posed for the camera were nearly nonexistent except in the case of minorities, who were photographed specifically to showcase their ethnicity and document their participation in the project. Moreover, none of these images were published but were held in reserve by the Bureau of Reclamation and the Bureau of Indian Affairs.

Although only a few photographs of Native Americans at the dam site exist, there were also a small number of illustrated depictions of native peoples. One example is from an organization called the Boulder Dam Science Company, which published a booklet titled *The Romance of Concrete and Steel* in 1933, in which we see a remarkable interplay of word and image. On the cover of the booklet is a large-scale Native American peering down onto the dam (figure 5.14). This presents a reversal of the typical imagery we have

Figure 5.14. "The Romance of Concrete and Steel: Boulder Dam," 1933 (pamphlet, illustration by Frederick A. Eddy, Boulder Dam Science Company, Wayside Press). Courtesy of the Boulder City Museum and Historical Association.

seen thus far in which the human figure is presented as miniature against gigantic nature or sublime technology. The inside of the booklet, however, is filled with illustrations depicting scenes of the dam under construction, a panorama of Boulder City, a topographical map of the Southwest, technical sketches of Boulder Dam's apparatus works, and more.

In the cover image the human figure is nearly naked and set alongside the great technological achievement. If it were not for the figure's tremendous size, his nakedness would suggest vulnerability or even helplessness. His attire also implies a primitive state of culture, one that is being altered by the technology. The figure of the Native American is clearly a romantic one, as is suggested by the title of the booklet, but what is also romanticized here is the technological achievement and the materials of modern construction—concrete and steel. The Native American figure is part of the old Wild West romance myth that is so important to the ethos of the American West. But the native also seems to be looking at the dam in wonderment and perhaps resignation. The figure appears to be looking at the dam as one looks at a curiosity. Although the native carries his own technology, an instrument of war and hunting, his bow and arrow are clearly impotent against the modern dam.[71]

Another image, this one from the book *Hoover Dam Including the Story of the Turbulent Colorado River* published the previous year, also shows a Native American examining the dam, as the canyon walls rise high above him and the large blue Lake Mead extends into the background. Two cars drive across the top of the dam, and the spillways are open, with a large volume of water spraying out into the river. In this image, we do not see the Native American's face, only his back. Again, the figure is nearly naked, with only a loincloth draped across his waist.

In these images, the old and new are powerfully contrasted. Native figures in traditional dress—feathers in their hair, and loincloths around their waists—look out onto the ultramodern dam, a symbol of industrial and governmental power, symbolism echoed in the oversized stately building placed at the bottom of the dam. Here we see that the native people are left behind, a romantic legacy of a time long past. This exclusion from participation crossed over into the realities of the dam's construction.

Every romance has a tragic figure, and here the Native American is clearly that character. In the early 1930s technology did not yet dominate the entirety of the West, just as industrial products do not dominate the

Romance image, but as is suggested in the *Hoover Dam* image, it will eventually happen. Beginning with John Wesley Powell's adventures down the Colorado River, and then with the settlement of the Imperial Valley and the Owens Valley and the building of Hoover Dam, Parker Dam, Grand Coulee Dam, and myriad other endeavors, the West was slowly being conquered by water reclamation projects. In these two images, Native Americans look out on the dam as a peculiarity, but soon the land and their way of life would be forever altered. Dorothy Hogner in the June 1941 edition of the *American Girl* (published by the Girl Scouts of America) offers an unusually poignant reflection (for the time) on the effects of the dam on the native people: "As we raced the motorboat over the imprisoned Colorado, we were suddenly aware of the violent changes machines have brought to our lives. Here, little cliff Indians once lived their simple existence, climbed like monkeys up and down the canyon sides, hid their meager crops in storage bins in the rock crevices, and made the old walls resound to their cries. But those echoes now are choked forever under the gurgling waters which have risen to claim the canyon at the command of machine-aged man."[72]

The drive to modernize savages, natives, and "lesser" nations generally is rooted in a sort of European social Darwinism devoted to converting the uncivilized to Christianity, a notion that closely parallels the drive to reclaim the West and tame the menacing Colorado River. And while native peoples did have intimate knowledge of the desert ecosystem and did labor to change parts of it to their agricultural advantage, their aim was not the complete domination of the natural landscape. Instead, they themselves, like the landscape, came to be dominated by the American mythos of modernity, one more step in a long march to pacify, civilize, and assimilate native peoples into American society. Whether this occurred through representational fidelity or through fictional compositions did not matter. Widespread depictions of primitiveness undoubtedly invigorated white Americans to exercise multiple forms of power over native peoples, ranging from scientific racism, to the ecumenical demands of missionaries for religious conversion, to the removal of native peoples from lands desired by the US government.

ALTHOUGH PHOTOGRAPHS only capture partial historic truth, their undeniable indexicality does serve a political and disciplining function that cannot be overlooked. Thus, the lack of a visual record from events such as labor strikes and industrial accidents widely known to have happened provides validity to the state's success in ensuring their exclusion.

Images of the dam were created and distributed primarily as a function of the state; thus, images play a crucial role in representing who is and is not worthy of citizen-state participation. The state establishes the relationship between the citizen and itself; therefore, the state's choices of what to not depict are as telling as its choices of what they do depict. In the visual story of Hoover Dam's construction, black citizens, union strikers, industrial accident victims, women, native peoples, and the dead are all largely excluded from the emotive potentiality of its photographs and, consequently, the possibility of future viewers to engage in their remembrances. This fact also begs a larger philosophical question: what is to come of the capacity of depiction to capture traces of time and place if the state decides that some citizens are not worthy of that honor while others are? In this way, preferred citizens are placed at odds with disenfranchised citizens through the unseen.

Photography has the ability to stand as a record of the past by eternalizing the present for those in the future. Images resurrected from the dustbin of history can challenge the ideological agendas of the state, while at the same time resisting the cultural milieu in which they were created. Far from being mere nostalgia, these images present something of a representational archeology, holding their form of knowledge making through eyewitness testimonials of what was there.

Although many women, blacks, and Native Americans lived and worked at the dam site, it is undoubtedly the dam itself and the machines and men that built it that dominate its presence in our cultural imaginary. This discussion serves to highlight the effectiveness of state-sanctioned efforts to control the depiction of certain peoples and to limit the circulation of images that might unveil undesirable cultural taboos or counter-narratives. However, these efforts, part of a larger political and social effort to control all visual and verbal testimonials of the dam's construction, were not restricted to minority groups but were employed in depicting class stratifications of whites as well. White laborers were not immune from the power of visual depiction to constitute their identities and social status. It is to this group of people that we turn in the next chapter.

6

Imaging Labor

In an article titled "Building the Big Dam" published in the June 1935 edition of *Harper's*, Theo White poignantly describes the construction project in terms of labor and the human toll the project took on those who chose to stay and work. According to him, manpower was the most pressing need in constructing the dam. "Man was the unit of work, not the kilowatt. Anything with legs and arms, formidable and hairy, was accepted," he remarked. However captivating his description, it was not truly the case, as many hundreds of minority workers who applied for jobs at the dam were denied because of vague "experience" requirements, or preferences given to white military veterans. Others were explicitly banned by language in the government contract. What Theo White does accurately represent—in disturbing yet poetic detail—are the wretched conditions that laborers who did make it onto the payrolls were forced to endure for the sake of employment.

After days of travel they were dumped at the camp, a spot remote from civilization. They were driven to work on scanty food of outrageous quality, noisome water, and venomous liquor. The weak died, or deserted and died. . . . They were robbed by exorbitant prices at the company commissary. And they were fortunate to show a penny at the completion of the project. In the end the weaker and licked quit to roam the streets of the metropolitan areas. Beggars or worse, they were contaminated mentally, morally and physically by the camp. The strong survived and went on to the next. The camps were wretched holes erected in the name of civilization that they might add some necessary appendage to civilization's needs—dam, railroad, or bridge. There was neither the virtue nor the cleanliness of the primitive existence.[1]

Although the dam project held the prospect of a new beginning in the midst of the Great Depression, the migrants who descended on the Las Vegas area and construction site found many of the same hardships they were trying to escape. Because of the worsening national economic situation, President Herbert Hoover and Secretary of the Interior Ray Wilbur urged Bureau of Reclamation commissioner Elwood Mead to begin construction on the dam six months ahead of schedule with the hopes of providing some job relief (and to thereby secure some "relief" for their own political futures). However, in a September 1931 article in the *New Republic* Edmund Wilson noted that as a result of the administration's desire to address (or *appear* to address) unemployment by endeavoring to start work on the dam as soon as possible, "the men are supposed to work in Black Canyon before proper facilities for living there and conducting work have been provided."[2] Regardless of the construction site's ill-preparedness, as word of the project spread, thousands of out-of-work men and their families began streaming into the area.

After the announcement of the construction timetable, jobless men and their families were arriving in southern Nevada in numbers rivaling those of the Comstock Lode (the first major US deposit of silver ore discovered in 1859 under what is now Virginia City, Nevada) when prospectors swarmed the area to stake their claims. Leo Dunbar, a government hydrographer who worked on the dam, recalls, "You've got to realize that when people found out that the project was going to be built, they flocked here from all parts of the United States." Dunbar continued that, with the dam being built at the height of the Great Depression, "the fact that there was going to be work here induced people to come in here—not only the men, but their families came with them. They just picked up whatever they had and loaded it into a truck and drove here, and had hopes of getting a job."[3]

The US Labor Department and State of Nevada employment office in Las Vegas reported that within days of initiating the process of hiring for the project, twenty-four hundred job applications and twelve thousand letters of inquiry had flooded their office. By 1931 that number would swell to more than ninety thousand job applications for fewer than six thousand positions.[4] The peak of employment at the dam was 5,250; however, less than 50 percent of those applying at the Las Vegas employment office received jobs because of limited openings, discriminatory hiring practices, and government provisions giving preferences to military veterans. According to the December 1934 edition of *American Legion Monthly*, 92 percent of (white) military veterans who applied received jobs at the dam.

But since people looking for work were arriving in numbers far outstripping the positions available, hundreds of people per day were also inundating the only soup kitchen open in Las Vegas. Hundreds more, including women and children, were going hungry. Although the government issued numerous statements that construction on the dam would not begin for months and that very few jobs would be available, people were not deterred. Wheezing automobiles and weather-beaten railroad boxcars delivered scores of people, while others crossed the desert by horseback and some even walked, all with the hopes of finding employment at the dam.[5]

Ragtown

Many of those who arrived in the area stayed in Las Vegas, but others went a few miles north directly to the site where construction on the dam was just beginning. People arriving at the construction site were greeted by a large squatter community spread out along the riverbank of the Colorado, an area officially named "Williamsville" but nicknamed "Ragtown" by those that lived there. The appearance of Ragtown was similar to the mid-nineteenth-century prospector and mining towns sprinkled across Nevada, such as those photographed by T. H. O'Sullivan. Like O'Sullivan's well-known image "Sugar Loaf, Washoe, near Virginia City" (1868), the photographs of Ragtown—with the dark shadows created by the tents and shacks set against the light background of sand and dirt, with the straight edges of the roof and canopy lines set against the sagebrush, dusty landscape, and rounded hills in the background—are reminders that this was the beginning of the Nevada landscape succumbing to industrial encroachment. Just a few months earlier, this area had been devoid of human activity. These early photographs of Ragtown and other "squatter" camps, the labor barracks built by Six Companies, and the dusty tracts of buildings in Las Vegas document the start of the vast transformation of the landscape and prefigure the population boom of Las Vegas and the construction of the dam.[6] But instead of being the utopia that some envisioned for the future sites of Las Vegas and Boulder City, these images depict the area surrounding the dam as becoming laden with the same social and cultural ills of the industrial revolution that once were restricted to cities and urban centers.

Erma Gobdey and her husband, Tom, who had moved from Silverton, Colorado, where Tom lost his job working as a miner, recalled the squatter town as being built mostly of "pasteboard cartons" or anything else that people could find lying around. "It was called Ragtown," Erma said, "but it was officially Williamsville. . . . Everybody had come in just a car with no furniture or anything. All we could do was put our mattresses down on the ground and get ready to go to sleep. Some had tents, but a lot just had canvas or blankets or anything that they could have for shelter."[7] Winthrop Davis also remembered the depressed state of the camp and those who lived there, writing that there were itinerant camps all around Las Vegas, what most people at the time referred to as "Hoovervilles," mockingly named after President Hoover as a symbol of his stern refusal to use government resources to stem the depression. In a section of his unpublished autobiography in which he discusses his time working as a photographer at Hoover Dam, Davis cautions that at the time there was no such thing as food stamps or government assistance of any kind that might have provided relief to the poor or unemployed. "People were living in cardboard shanties," he recalls of the early days at the construction site, "and those that died were buried in the potters field. I remember the stone heart that Vegas once had when they shut off the tap water in the cemetery where most of those poor hungry people got their water."[8] And though it was called Williamsville and nicknamed Ragtown, Joseph Stevens observed that "an even more accurate name would have been Cardboardtown, for that was the principal material used in building the hovels that most of the residents called home. Scrap lumber, flattened oil and gasoline cans, tarpaper, and burlap were also used to fashion crude structures that provided at least some shelter from sun and rain."[9]

W. A. Whynn, in his unpublished memoirs of his time at Hoover Dam, wrote that what existed in the ramshackle huts of sticks, carpets, cardboard, sheets, and rags that protected the few belongings that families brought on their journeys to the dam site formed a "village" compared to which "the slums couldn't begin to hold a candle."[10] He recalls that of those who lived in Ragtown, "a big majority of them were broken or badly bent and didn't have anything to live with or any way of buying life's needy [sic] luxuries." Entire families lived under cardboard roofs of only a few square feet. There were "no toilets or bath tubs or any streamlined conveniences to amount to anything," Whynn wrote. "A whole family, kids, dog, cat, parents and what not crowded into these make believe houses. And I don't lie when I tell you readers they were make believe houses." He describes people who were the most pitiable of paupers eking out desperate lives in the middle of the Mojave Desert with only the hope of finding a job at the dam once construction started. So destitute were these individuals and families that, according to Whynn, they "really and truly didn't have a saucer of their own."[11]

Once Six Companies finished construction on the barracks for laborers and their families, those living in Ragtown were encouraged to move, but many stayed put, complaining that the buildings were thrown up so quickly and built so shoddily that holes were left in the roofs and gaping cracks under the windows and doors let in rain, heat, mice, and especially the fine dust of the desert whipped up by the ever-present winds, a dust that would inundate everything, including the food. According to Whynn, a common saying was "We eat a peck of dirt before we die."

Although these kinds of stories recounting the desperation of those seeking a new life at the dam are abundant, the adversity faced by the earliest settlers is best expressed in photographs that also serve to highlight the important differences between visual and verbal representations in allowing a full appreciation of their meaning. One poignant example can be seen in a series of similarly sited photographs from August of 1931 titled, simply, "Williamsville." In one of the images (figure 6.01) we see three adult women, two in the foreground posing and looking at the camera, and one barely noticeable in the background whose face is hidden by a hanging piece of clothing and who does not engage the camera. In front of the women are four children, none with shoes even though they are walking on sand, dirt, and rocks. In fact, the only person in the photo who is wearing shoes is the woman in the background. The children are rather unkempt, with dirt strafing their legs and smudging their faces and clothes. The structure under which they're standing is built of a few pieces of wood over which is hung a tarp. There are clothes hanging from the structure and from the clotheslines tied to the trees, while a galvanized clothes-washing bin hides under what appears to be a bed frame on the right side of the structure. The floor is made of wooden planks, while the support beams along the bottom are held off of the ground by a motley collection of wood and rocks. But the structure has no walls to provide protection from the elements; the floor serves merely as a mooring for the posts and a demarcation of space rather than any kind of

Figure 6.01. "Williamsville," August 13, 1931 (photographer unknown). Courtesy of the Boulder City Museum and Historical Association.

domicile luxury. A second image is of the same structure but showing only children.

These images exemplify a communicative plane of visual representation —an immediate and visceral response—that words cannot replicate. The framing of the photos suggests that what we see encompasses the entirety of these people's possessions. Everything in these images, from the people to their meager structure to the framing of the photo, emanates a feeling of *poverty*—a pauperism so deep, a hardship so immediately evident that it instantly conjures a sense of foreboding that few of us have ever personally experienced yet can somehow appreciate by viewing the photograph. One can write the words "poverty stricken," but those words do not convey the immediacy of expression provoked by viewing the hungry faces, dirty children, uneasy stances, or looks of resigned destitution that the image can produce.

Although the "Williamsville" image is now often selected for its ability to communicate the hardships of those living at the dam and to bring into stark relief why the name "Ragtown" was so appropriate, these photographs were not in circulation at the time they were taken. As far as is known, these photos did not appear in a single newspaper article or official government document from the 1930s, 1940s, or 1950s. This lack of circulation, again, reinforces the notion that the Bureau of Reclamation was highly successful in sequestering images that would or could contest their desired narrative of thousands of gainfully employed people, happily working at the dam. Moreover, this image is also particularly striking when viewed in contrast to the lavish-by-comparison homes built for the company executives and their guests. For example, one undated government promotional booklet exults in the "Spanish-style stucco" homes that were built for directors of Six Companies and their guests. This home, with its "veranda, fireplaces, and sweeping views," exemplifies the wide chasm between the contractors, engineers, and government officials and their guests and the poor working-class laborers of Ragtown (figure 6.02).

Another photograph, this one taken by Winthrop Davis and titled "Father with Two Small Children" (figure 6.03), depicts a man with two children milking a burro in an open field of sagebrush. The burro provides a dark

Figure 6.02. "Interior of Six Companies, Inc. guest house," April 10, 1932 (photographer unknown). Courtesy of the University of Nevada, Las Vegas Libraries, Special Collections. Burrell C. Lawton Collection.

Figure 6.03. "Father with Two Small Children," ca. 1930-31 (photograph, Winthrop A. Davis). Courtesy of the Special Collections, University Libraries, University of Nevada Las Vegas, Dennis McBride Collection.

contrast against which we can clearly see the figures, with no other contextual cues in the frame. The chained animal appears forlorn, as there are large patches of fur missing from its back and its ears are turned back and its head tilted down in something of a calm resignation. The two shoeless children accompanying the man look eagerly and intently at the discharge of milk into the metal pail. The man holds the younger child against his body with his left arm while he milks the burrow into the pail held between his knees with his right. Both children look on expectantly, with their unwary curiosity and white-blonde tufts of hair intensifying our sense of their innocence. In this image, we also cannot see any of their faces; however, since the man's face is hidden behind his toddler, it suggests a self-conscious loss of masculine pride in suffering the indignity of joblessness coupled with a daily struggle to feed his children.

What is remarkable about this particular photo, however, when considered in the context of the entirety of Hoover Dam imagery, is that just as women were rarely depicted with technology, this is the only Hoover Dam–related image I have seen that depicts a man doing what was accepted as the woman's job of tending to children. The scene in the photo, considered within the framework of the prevailing homogenization of male versus female depiction (men with technology and women with children), hints at a possible reason as to why we see what we do. But it is not until Davis provides a verbal narrative of his encounter with the man in the photo that we get a sense of the full meaning of the image. In an oral history interview with Patricia Lappin for the Boulder City Museum and Historical Association, Davis said of the photo, "A lot of people died from lack of nutrition and dehydration and a little of everything. You see that picture in the museum in Vegas and there's a picture of a young man milking a burro. When I took that picture, he said, 'This is all I got. This milk and a bag of corn meal is all we have.' He said, 'Pray for me brother.' [The children] happened to be twins; his wife had just died a few months before that from lack of medical attention."

Unlike in the "Williamsville" photographs, the more impactful meaning and heartrending emotion emanate not from the image alone but from Davis's verbal account of the photo. By filling in details with his narration of the photograph, Davis is able to provide a more robust context, thus enhancing the meaning of the image in a way that would elude even the most perspicacious "reading" of the image, and underscoring that full meaning can sometimes be produced only through a synergy of the visual and verbal.

The Most Inhospitable Land on Earth

Winthrop A. Davis, who arrived in Las Vegas in 1930 with the intent of photographing and writing about the people and events in the still-sleepy city of Las Vegas and the construction of the dam, remembers vividly "drinking alkali water and being basted in the desert sun while photographing some of the most inhospitable land on earth."[12] He remembers the early days of the dam's construction in which people had difficulties meeting even the basic necessities of life such as finding drinkable water. "The times were bad," he said, "I don't know how they could have been much worse, when people can't eat, can't get medical. So many of these people didn't even have good water to drink, potable water. They drank it out of the ditches, they drank it out of the river, and that river, I know, I drank plenty of it. I got awful sick from it. So much alkaline in it my shoes would turn white where my sweat would come through my shoes."

In his unpublished autobiography, Davis recounts with a sense of bitterness and despondency what he felt was a lack of respect that modern-day Las Vegans have for those who built the dam. To him, the dam stands as a symbol of inequity and suffering, while the travails of laborers and their families are discarded as a bygone memory glossed over by the wealth and decadence that Las Vegas stands for today. "When I go to Vegas now, I am not titillated with all their neon signs and their obstreperous trappings, for I can still remember seeing those undernourished men, women, and children living amongst the mesquite trees in their cardboard shacks, many of the men who fought in the first World War while the Duponts, Morgans, and others got rich, were now just asking for a job that they might feed their loved ones. I hope I never have to witness such an experience again."[13]

It was experiences like these that shaped Davis's view of the Hoover Dam project and intensified his respect for the people who endured its incredible hardships while holding out hope of finding backbreaking, often life-threatening work that only paid between $4 and $5 per day with no benefits. Unlike other photographers such as Ben Glaha who celebrated the dam's machine aesthetic, Davis said his "greatest interest was in the people, how it affected them, their aspirations, their dreams and how they were treated." In his letters, memoirs, and interviews regarding his time at the dam, Davis always returns to the hardships he witnessed and the people he met. In a 1995 letter to Dennis McBride, Davis lamented the schism between what he described as "a monument to two different worlds: one the public sees, which is a marvel in engineering; the other one is the greed and corruption and disregard for the lives of those who built it. There's time when I feel I knew too much and saw too much, and much I would like to forget."

Although Davis took many photographs of the project, and for a time made a living selling them to various publications, he quit photographing the project when it was roughly two-thirds complete because to him the magazines and newspapers were only concerned with the engineering and technological challenges of building the dam, not its human stories.[14] When the technical challenges were met, there was little saleable documentation left to do.

The Persistence of Heat and Death

The difficulties posed by the persistent lack of food, shelter, and drinking water were compounded by the crushing heat of the desert, something that nearly every memoir and oral history by those who lived and worked at the dam recounts in stark detail. Yet heat is a feature of the natural environment that is particularly resistant to photographic depiction, thus leaving us with verbal descriptions as the principal way through which we can appreciate its effects on those who lived and worked at the site, where the daytime temperatures in summer were likely to soar above 120 degrees and barely dip into the nineties at night.

Journalists, engineers, and laborers alike describe the heat as suffocating and relentless. Edmund Wilson remarked in a 1931 *New Republic* article that "even the government engineers who were the pioneers at Black Canyon describe it as 'Hells Hole' where you get 'goofy with heat.'" Wilson goes on to describe the canyon walls as "cinders which never cool off: they might still be smoldering from the volcanic disturbance which has left the whole landscape an infernal desert. You can't touch them without getting scorched and the very winds that blow through them are furnace breaths. The temperature in summer is often 123 in the shade."[15] Erma Gobdey also recalls the heat: "It was terrifically hot. My God, it was terribly hot and dusty. . . . It would get to be 120 by nine in the morning, and it wouldn't get to be below 120 before nine at night. . . . I would wrap my babies in wet sheets just so they could sleep."[16]

Winthrop Davis described how the geologic black basalt rock of the canyon exacerbated the high temperatures by absorbing the heat of the sun all day and releasing it at night, which provided no respite from the omnipresent burden of the heat. "Even on the pier we were building it would get to be 105 degrees at midnight, which made sleep difficult," he recalls. "We would take our sheet and walk down to the river naked and, after thoroughly wetting the sheet and our bodies, we would wrap the sheet loosely around us and then lay down and try to get sleep before the sheet dried out."[17] Edmund Wilson similarly described how "the nights were so suffocating and restless that [the laborers] welcomed the diversion of work," and that the heat was not the only irritant. "The camp is infested with some kind of wild mice," he wrote, "which jump up onto the beds with incredible agility and crawl all over the men." Eventually the men got so used to the mice crawling on them at night that "they brushed them away in their sleep like flies."[18]

Death, too, was seemingly on the minds of many people working and living at the dam site. Winthrop Davis, who had been working as a boat concessioner taking survey parties, government officials, and engineers up and down the river in the early days of the project, wrote that once his photography started to sell to a few magazines and newspapers, he was ready for his girlfriend, Elsie, to move out to Las Vegas, where they were to be married. Upon her arrival, Elise told him of the disheartening circumstances she had witnessed on her journey. "Elsie had arrived and not in the highest of spirits," he wrote, "for she has seen the highways spotted with ragged, hungry people, walking [along the burm] with shoes made of burlap sacks wired onto their feet, and at each inquisitive filling station [an] agent who inquired where she was going while pumping the gas would say: 'There's people starving to death there.' The truth of the matter, there was."[19]

Bruce Bliven in a 1935 *New Republic* article recounts a story by Buddy, his chauffeur around the dam site, who described the way in which people would cross the river by traversing a twenty-four-inch-wide cable bridge with no handrail that was strung seventy feet above the river. "Many a man came here, took one look at that bridge and quit," he said. Occasionally people would lose their nerve and "drop to all fours, hugging them planks, and crawl off an inch at a time." In order for two people to pass each other they would engage in an aerial dance in which "you'd take hold of hands, swing out and past each other, straighten up and let go of hands again." It wasn't until someone engaged in this midair pirouette got scared in the middle of the "swing" and let go, causing the other person to fall to his death, that the Bureau of Reclamation strung another cable as a handrail. Unfortunately, according to Buddy, it "wasn't much good, though; kept hitting you in the leg as you walked and you'd come off of there black and blue."[20]

As these remarks imply, death marks much of the narrative of Hoover Dam. "The weak died, or deserted and died," Theo White wrote of those who worked at the dam. One bronze plaque at the dam, dedicated on May 30, 1935, reads "They Labored That Millions Might See a Brighter Day—In Memory of Our Fellowmen Who Lost Their Lives on the Construction of This Dam."[21] However, these plaques did not satisfy many people. Winthrop Davis complained in a letter to the Nevada State Museum, "All I saw upon completion of the dam was a large plaque which read, 'Built by the Reclamation Bureau and Six Companies Inc.' No credit was given to the 15 or 20 thousand men who gave five years of their life time, and some gave their lives as well."[22]

Death came in myriad ways: some men fell from the canyon walls, others fell from the dam; some were electrocuted, others drowned in the river; some died of heat prostration, others of falling rocks, debris, or tools; some men died in truck accidents, while others died in explosions. A display at the Hoover Dam visitor center shows "quotes from actual obituaries" that describe in sometimes macabre detail the final moments of several workers: "Dam worker crushed to death under scores of tons of muck," or "A falling jumbo yesterday crushed every spark of life from the body of Victor K. Auchard, 25," and "G. D. Isaac, rigger . . . met almost instant death by electrocution." The display also states, "Throughout the job site, billboards reminded workers that 'Death is So Permanent—Be Careful.'" W. A. Whynn describes the "special deliveries," a nickname laborers gave to the ambulances that would regularly zoom up the canyon roads "headed for the hospital with some poor unfortunate." He recalled, "At the time [the ambulances] were kept awful busy. . . . Many a poor man took his last ride up the canyon in one of those specials."[23]

There are, however, conflicting tallies of people who perished there. The official number of deaths stands at 96, but if starting from the earliest surveying trips, the number rises to 114. Many people have argued that carbon monoxide poisoning was the leading cause of death, but since nonindustrial accidents did not trigger insurance benefits payments, government records show forty-two "pneumonia"-related fatalities, presumably listed as such to avoid insurance and worker's compensation claims.[24] Likewise, in keeping with the government's practice of limiting as many claims as possible, a wife or widow, for example, could not file a workman's compensation insurance claim for a husband who had been killed or injured on his day off as he was not officially "on the job" at the employ of the dam or government on that particular day.[25]

Remembering those who died is an important part of the cultural memory of the dam, but two tales of death in particular have persisted since the 1930s, likely as a means to further humanize or dramatize the project. The first is that there are men buried in the dam. As any engineer would attest, a body buried in the concrete of the dam would be a significant safety hazard, as the pocket of air left from a decomposed body would become a point of structural weakness. Although many a jokester planted an empty pair of rubber boots into still-wet concrete, Daniel Rosenberg's article "No One Is Buried in Hoover Dam" attests that this story is a complete fable, yet it is one that still endures despite the Bureau of Reclamation's best efforts to debunk it.[26]

The second is the oft-repeated story of the arc of deaths at the dam that started and ended with two men—a father and son. The received historical account is that on December 20, 1922, surveyor J. G. Tierney was the first person to die at the site when he drowned while scouting for a location for the dam, and that precisely thirteen years later to the day, his son, Patrick W. Tierney, was the last to perish while working on the dam. Although this poignant yet morbidly ironic story has been frequently repeated, I question its veracity. For example, on May 28, 1921, an article titled "Youth Loses Life in Boulder Canyon" appeared in the *Las Vegas Age* of Clark County, Nevada, reporting that another man, twenty-year-old Harold Connelly, who had been "working for the government at the site of the proposed dam in Boulder Canyon since last December," had drown on May 15—nearly *two years* before Mr. Tierney's death—while swimming "at the point frequented daily by the government employees engaged in the work there." Additionally, the article begins with the statement "The treacherous stream of the Colorado claimed *another* victim" (my emphasis), suggesting that Mr. Connelly was not the first person to perish at the proposed dam site (though I have found no documented references to deaths of government employees or contractors working at the dam site before this). The question for the historical record, I suppose, is whether Mr. Connelly was "officially" working on the dam project, merely working for the government at the site of the proposed dam, or took his unfortunate swim on his day off or while "off the clock." The article does not specify. Moreover, considering the large number of deaths linked to Hoover Dam's construction, it is hard to imagine that nearly two years passed without a single fatality until J. G. Tierney's unfortunate accident. These types of ambiguities make it especially difficult to render an exact tally of deaths at Hoover Dam and lend strong credibility to assertions that many more people died in the service of its construction than have been officially reported.

Memorializing the Dead

In the late 1930s, an effort to commemorate contributors to the dam project caused a controversy that reached the highest levels of government. An editorial in the March 22, 1937, *Las Vegas Evening Review-Journal* questions how some names were selected to be included on a plaque commemorating the dead while others were not. The article, "Honor Where Honor Is Due," rebukes Secretary of State Harold Ickes for having taken it upon himself to place a select few names on a plaque to be installed at the dam. According to the piece, "If you were to base your views on the information contained in the plaques designed by the Secretary, you would gather that nobody played a part in the project except the Bureau of Reclamation." The article claims that all but seven names are from either the bureau or one of the Six Companies contractors. Senators Hiram Johnson and Phil Swing, too, are mentioned, but no governors, no state engineers, no journalists, and no community members. There is also no mention of the decades-long struggle to ratify the Colorado Compact or those who worked to see its completion. "Most importantly," the article continues, "what glory is there for the men who actually did the work of building the dam project? The muckers, shovel-runners, concrete workers, carpenters, and the rest. Strangely, they had to DIE to gain any recognition. The names of those who lost their lives on the project were the first to be engraved on the mighty structure. But the other thousands who were just as necessary a part of the program . . . must gain their glory from the knowledge down in their own hearts that they contributed their bit to the control of the once savage river."

Secretary Ickes's commemorative plaque drew even more ire from Senator Key Pittman of Nevada, whose name was also not among those selected. In the previous week, the *Las Vegas Evening Review-Journal* reported that Pittman and Ickes "had argued the question in 18 months of correspondence,"[27] while the March 30 edition ran an article with the headline "Senator Irked at Secretary's Honorary Ideas: Pittman Lodges Protest as Only Two Nevadans Put on Plaque." Pittman's plan was to establish an architectural and memorial commission to oversee the process of naming and design, but Ickes dismissed this as "unnecessary" since the Department of the Interior "has already completed memorializing plans." The editorial ends with the unwieldy conclusion that plaques are inadequate, but if plaques must be used, then they should include everyone who contributed, though the practicalities of how one might immortalize the names of every contributor to the largest engineering project in the history of the country, one that took decades to plan and execute, employed tens of thousands of people over the course of its construction, and included negotiations between every state in the Southwest, are not mentioned.

The resolution to the question of how to commemorate all of those who contributed to the project was eventually established through the use of symbolic representation, which is abundant at the dam. Norwegian-born, naturalized American citizen Oskar J. W. Hansen, who designed much of the artwork at the dam, recognized the importance of representing everyone, not just the powerful or moneyed. He wrote that his work must deal "with many intangible values" and submits that "if this were not so then men would not commemorate their existence in such monuments. The problems which the sculptor must face and with which he must deal in his art are the problems analogous to the infinite paradox of life itself."[28]

In keeping with the approach of New Deal artwork, Hansen glorified the laborer.[29] One of his pieces at the dam is a bronze plaque commemorating the men who died while working on the project with the words "They died to make the desert bloom" arcing across the center of the piece. In this image a laborer rests with his legs beneath the waves, his impossibly narrow torso leads up to a broad, muscular chest and shoulders, his face turned upward toward the sun, while the symbols of power, irrigation, and conservation rise above him. The hands of the laborer are stretched upward, suggesting homage to the Great Spirit and to the power required to achieve feats such as building the dam and controlling nature. At the top of the plaque are wheat and gourds, food staples of indigenous peoples that also symbolize the bounty that will result from the continuous supply of irrigation water to the Southwest. These images are described by one writer as expressing "the artist's belief that productive control over natural forces is obtained through trained physical strength and knowledge of our environment."[30]

On the Nevada side of the dam is a 142-foot-high flagpole flanked by two thirty-foot-high winged sculptures, which Richard Guy Wilson notes were "surrealistic apparitions" that underscored the otherworldly nature of a dam and lake in the middle of a forbidding desert.[31] The entire dam site includes a total of ten bas-reliefs designed by Hansen: five depicting the purposes of the dam (flood control, navigation, irrigation, water storage, and power), and five representations of the native tribes of the Southwest, with the inscription "Since primordial times, American Indian tribes and Nations lifted their hands to the Great Spirit from these ranges and plains. We now with them in peace buildeth again a Nation."

Images of Laborers

During Hoover Dam's construction, Union Oil, the Bureau of Reclamation, the Metropolitan Water District of Los Angles, *American Steel & Wire*, *Construction Methods* magazine, *Fortune* magazine, the *Engineering News Record*, *Western Construction News and Highways Builder*, *Compressed Air Magazine*, and many other publications circulated heroic accounts of the dam as a sublime technological triumph. However, these celebratory accounts of the engineering and construction processes often overlooked the laborers who were building the dam. The mobilization to construct the dam took years, and for laborers and their families it required overcoming immensely difficult living and working conditions. In images of laborers employed or seeking employment at the dam, we see them typically presented either as heroic or as the disenfranchised underclass. Two schools of artistic thought that emerged during the Depression era, *regionalism* and *social realism*, each of which held different assumptions about and represented contrasting artistic perspectives on the era, prove helpful when considering images of Hoover Dam laborers.

Regionalism, described by some as "anti-colonialist art," was popular in America during the Depression, reaching the peak of its status from 1930 to 1935.[32] Notable regionalist artists such as Grant Wood, John Steuart Curry, and Thomas Hart Benton painted images that not only reflected the events of the 1930s but also attempted to portray the national mood as they saw it.[33] At a time when the American public was looking for anything positive to lift its spirits, those with a regionalist perspective tried to express in their art the values of hard work, self-reliance, and community. They also sought to preserve the Jeffersonian ideals of agrarianism, the independent farmer, the small-scale entrepreneur, and the principles of small-town life on which, in their view, America was founded.[34] Regionalism visually contrasted a nostalgic past with a moribund present to show the cold standardization imposed by industrialism.[35] Despite its name, regionalist artists aligned themselves with federal New Deal initiatives and aimed to contribute to the formation of a national culture through imagery that exuded the ethos of the "manly man" living the American dream through hard work.

Social realism was another art movement of the Great Depression that attempted to depict the working-class as heroic; however, it developed as a backlash against romanticized and idealized notions of American labor.[36] As the ills of the Industrial Revolution became apparent—the proliferation of urban centers and slums contrasting with the wealth of the upper classes—social realist artists depicted the ugly realities of contemporary life, the racial injustices of society, and the low living standards of working-class people and the poor. Instead of seeing American culture as a source of inspiration, social realists saw injustices perpetuated upon workers in the factories and thus argued that New Deal programs alone were insufficient.

Though they represented several different and often radical persuasions—anarchists, communists, socialists, Trotskyites, and so forth—social realists shared a collectivist perspective. They saw art as a means to display the "unvarnished truth," and as a "weapon" capable of bringing about social change. Although social realists fought to influence the masses through their works, the public was not particularly receptive to their shocking depictions.[37]

Images of Hoover Dam laborers, then, can be divided into two categories: the social realist–inspired images of the downtrodden, oppressed, working-class poor looking for work, or the regionalist-inspired idealized and romanticized masculine worker. The regionalist perspective depicts laborers as healthy, robust, well-muscled objects of sexual desire, in control of the technology placed at their disposal and optimistic about the future. In contrast, the social realist perspective shows laborers not as working-class heroes, but as the anonymous and undistinguished working poor, toiling at the behest of the state, and the foil to both natural and technological sublimity.

Each of the depictions I just mentioned also coincides with larger cultural shifts in Americans' notions of labor, technology, and engineering and reflects deep-rooted social anxieties about the future of labor—indeed of masculine identity itself—in the machine age. As photographers, artists, and journalists chronicled the dam's construction, and as publications reproduced images and articles about the project, these portrayals of human labor bore witness to both the difficulty of working at the dam and wider anxieties and tensions in early twentieth-century America regarding the role of labor in a modern society, anxieties that were being widely articulated in art of the New Deal era.

One prominent example of this notion was photographer Lewis Hine. Hine's images of child labor, which are some of the most recognized images of the 1920s and 1930s, appeared in publications across the country and helped to fuel the progressive movement.[38] Later, Hine's 1932 book *Men at Work* moved beyond progressivism and explored imagery of the American industrial worker and iconography of working-class men.[39] Unlike most Hoover Dam imagery that celebrated technology while de-emphasizing the significance of the worker, the tools and mechanical settings in Hine's *Men at Work* are secondary accoutrements to the work itself and to the iconicity of the worker.[40] Hine and other twentieth-century photographers, along with sculptors such as Max Kalish and artists such as Arthur G. Murphy, Leon Gilmour, Stanley Wood, and Frank Cassara, portrayed the laborer as a symbol of American moral integrity. Although the iconic imagery of the worker with bulging biceps so familiar in New Deal art did not originate in the 1930s, since such images were also prevalent in the late nineteenth century, such as in Thomas Pollock Anshuntz's *The Iron Workers Noontime* (1880) and Ferguson Weir's *Forging the Shaft* (1877), these types of images became widespread in the late 1920s and early 1930s when romanticized images of the chiseled bodies of male industrial workers became a central visual icon of American culture.[41]

Arthur G. Murphy was one artist who portrayed the worker-ideal as a hulking laborer. Murphy's "Bridge Worker" series of sketches depict bare-chested workers with exaggerated, almost cartoonish physiques— bulging biceps, broad shoulders, and rippling back muscles—the embodiment of what Barbara Melosh describes as the "manly worker ideal."[42] Murphy's representations of industrial labor as brawny, masculine men hammering and riveting suggests a heroic notion of the laborer at a time when, in fact, unparalleled numbers of workers were being laid off by industries of all kinds. Other notable works such as Leon Gilmour's woodcut *Cement Finishers* (1939) and Frank Cassara's lithograph *Drillers* (ca. 1939) also feature shirtless, muscular men doing physical labor. But although Gilmour's and Cassara's laborers symbolize the ethos of American work, the men in both images look depressed. They have their heads down, concentrating only on the job at hand, not on future possibilities. In these types of images, American artists also sidestepped important social labor issues such as power and control of industrialists over working conditions and compensation and instead centered on the body of the worker himself.[43]

Gilmour's and Cassara's images bear striking resemblance to many photographs of Hoover Dam laborers, who are also well muscled, shirtless, and alone in the image. In figure 6.04, for example, we see a laborer working a pneumatic drill against the rocks. His muscular body is the focus of the photograph, but this is not a celebratory image, as the worker's head is also down, concentrating on the job at hand. He does not engage the camera and does not smile, but instead has a look of resignation.

As mass production, powerful machines, and improved efficiency forced a decline in the need for labor, images like this one followed a trend in New Deal art that depicted workers as vulnerable. In these images, the laborers are metaphorical—they symbolize work, yet also embody deep-felt

Figure 6.04. "Driller," ca. 1932 (photograph, Bureau of Reclamation). Courtesy of the Boulder City Museum and Historical Association.

anxieties of nonworking males of the Great Depression era in which unemployment was at times near 25 percent of the working-able population. The shortage of work struck at the heart of American identity and aggravated long-standing tensions between notions of class, masculinity, and character. Because of increasing efficiency and machine technology, the physical prowess of the laborer was becoming superfluous. At the same time, the artistic romanticization of the mythologized laborer of the past was on the rise.

As the New Dealers strove to enact policies to put men back to work, other government agencies sponsored artwork that portrayed laborers as already hard at work in traditionally masculine jobs. Notions of manhood and womanhood figured prominently in government-sponsored public art, such as that of the Works Progress Administration's (WPA) Federal Art Project and the New Deal's Treasury Section of Fine Arts. New Dealers used the art both as inspirational images to lift the morale of Depression-era citizens and as propaganda for New Deal policies.[44]

One example of this in Hoover Dam imagery appears in the May 1934 edition of *Fortune* magazine, which featured a series of eleven stunning watercolor paintings of the construction site by artist Stanley Wood, courtesy of a sponsorship by the Public Works Art Project (PWPA), of which nearly all of the originals have been missing since the 1930s and are among the most sought-after artworks of the New Deal era (figure 6.05). According to the article, "Mr. Wood's appointment as 'state artist' on the Boulder Dam project was one of the PWPA's least conventional and most richly stimulating assignments." It describes Wood and other artists of the PWPA as knowing that this type of work "is no mere relief measure, but America's (almost the world's) first official recognition of the artist as a useful member of society." The article goes on to defend the arts projects, describing how, through ventures like the one gracing its pages, "democracy has turned its attention to the worthiness of its art and to the support of its artists. Here, in fact, is a challenge to the American artist to prove what needs no proving; his place and his worth in the world." Wood's watercolors pay homage to Hoover Dam laborers, whom the article describes as "the little army of men," the "buggish and casually heroic men who build out of ancient desolation so great a monument." One painting shows the control room of the cableway system, what the article deems the "Puppet Master," and it goes on to describe Wood's paintings as a "constant motif" of picturing the laborers engaged in a "deliberate aerial dance of men and buckets, steel and timbers dizzily strung above the chasm." Wood's paintings, along with the article's descriptions, portray man and machine working together to build the dam, likening the "construction stiff"—the anonymous high-scalers, drillers, and puddlers,—to unsung heroes who share a kinship with the Roman legions who built the aqueducts, roads, and walls that still stand today.

Here we also find the recurring theme of death connected to labor at the site. The caption "Suicide for Profit" given to one untitled painting of shirtless drillers excavating the walls of the canyon suggests that the work is highly dangerous, enabling corporations to make profits at the expense—the

Figure 6.05. Boulder Dam Agitator Going Down to Penstocks (Nev.) Powerhouse Abutments Below, February 11, 1934 (watercolor, Stanley Wood, painted under the Public Works of Art Project, reprinted as *Commuters: Downtown Trip, Fortune*, May 1934).

death—of laborers, because "so far, despite hard hats, a field hospital, and all precautions, eighty-two men . . . have given up their lives to their monument." The laborer is "America's actual builder and unsung hero," it continues, "and he's a man you'll hardly see the like of outside America in this time." Another painting was given the caption "Like Janus, the Dam Has Two

Faces," while the article describes the scene as a "fantastic mass of planes and terraces, which might be the Hanging Gardens or a Viennese apartment house." It goes on to vividly detail the work shown in the painting, describing the different functions of the dam that appear in the image, including the powerhouse, the inspection galleries, the chutes for discarded lumber used in the concrete forms, the three passengers in the skip descending onto the worksite, and, of course, the "little army of men advancing on the dam" for the 3:00 to 11:00 P.M. swing shift.

Seeking to capitalize on the ethos of masculinity and work in traditionally male-dominated jobs, advertisements in the 1930s also echoed this theme. For example, in the same edition of *Fortune* magazine, an advertisement for the Reading Iron Company features an image of a shirtless, brawny man looking on as two other men work ladles in an oven. This image is used to suggest "craftsmanship" over science as the reason Reading Iron's pipe has stood up "in more places, over a greater span of years, than does any other kind of pipe." The advertisement states, "In the stern work of making metal that defies time, all of man's science, all of his invention, are not enough. Something else is needed—CRAFTSMANSHIP." In another advertisement, this one in the September 7, 1931, edition of the *Las Vegas Evening Review-Journal* for Adock & Ronnow, a Las Vegas men's work and dress clothing store, a powerfully built, shirtless man holds a sledgehammer in front of a large cog. The advertisement celebrating Labor Day says, "Hats off to Labor: And their wonderful opportunity presented for service in the construction of the great Hoover Dam."

Although many images documenting Hoover Dam's construction depict the laborer in a way that reflects 1930s government-sponsored public art showcasing muscular, "manly worker" images that evoke the shared cultural assumptions of what it meant to be a man as expressed in New Deal art, some images also suggest a subtle yet curious dissonance in their presentation, one that appears to eschew the masculine of the manly worker for the traditionally feminine depiction of subjects in Western art. Although we have already seen how the dam's sublime technology was portrayed by way of contrast with tiny human figures, in these images it is the men, not machines, who elicit our reverence. In presenting the male body as a glorified icon of work, these images present laborers as objects of the (male) viewer's desire, a situation that seems contrary to traditional conventions of Western art, in which the viewer is presumed to be a male. As John Berger observes, "In the average European oil painting of the nude the principal protagonist is never painted. He is the spectator in front of the painting and he is assumed to be a man. Everything is addressed to him. Everything must appear to be a result of him being there. It is for him that the figures have assumed their nudity. But he, by definition, is a stranger—with his clothes still on."[45] In the dominant convention of Western art the female is either the symbol of idealized nature or the object of sexual desire.[46] If the female is to stand for nature, she is stripped of individuality and her face covered, shadowed, or limited in detail. If she is the object of sexual desire, the male voyeur objectifies the female, but her gaze does not engage the viewer. Berger describes this as "looking." To him "we only see what we look at. To look is an act of choice. As a result of this act, what we see is brought within our reach—though not necessarily within arm's reach."[47]

Kevin Barnhurst writes that a woman becomes the object of sexual desire in an image by being isolated, "taken unaware, while sleeping, looking off or turning her back to the spectator."[48] But he reminds us that the dominant depiction of the "natural" male figure is very different from that of the female in Western art. He writes that a male is nearly always clothed, posed upright and in control (only reclining in death), with his gaze either at the viewer or at a woman. He continues, "Instead of being a flowing unity, any exposed flesh is divided into panels or sections like armor. His genitals, frequently covered or hidden in darkness, are evoked by phallic objects or forms. And the surrounding ground presents other tokens of wealth, learning, or power."[49]

Nevertheless, in several photographs of Hoover Dam laborers we can see the male subjects in the image evoking *feminine* characteristics—the opposite of the "natural" male figures that Barnhurst describes. In these images, males are portrayed as both idealized nature and objects of sexual fantasy; they are, in fact, quite regularly eroticized. For example, the photo "Workman with a Water Bag" (figure 6.06), taken by Bureau of Reclamation photographer Ben Glaha, shows the laborer shirtless; however, his face is hidden as he takes a drink of water. This image has a singular focus on one unidentified individual's chiseled physique, an intentional act on the part of the photographer. Here Glaha purposefully sought to highlight the iconography of the worker over the sublimity of the machinery. But more importantly, this image and others like it go beyond mere "documentation" and move toward symbolism by incorporating formal aspects of iconography and the heroic worker. As Barbara Vilander notes, these images convey ideologies rather than mere facts and information.[50]

Glaha's iconographic "Workman with a Water Bag" works as an allegory (again, typically reserved for depictions of women in Western art) for the entire Hoover Dam project. It shows the laborer with no tools in his hands, only a water bag, suggesting the human element of labor enduring against nature without the need of machinery. The man's flesh contrasts with the sharp rocks and equipment surrounding him, while the short focal length of the camera renders the workers in the background out of focus and unidentifiable. The water alludes to the river itself, and thus the dam's raison d'être. The enormous dam is required to control the Colorado River and its destructive floods, yet the river will also provide drinking and irrigation water and electricity to the entire Southwest. The irony here is that the water is both

LEFT: *Figure 6.06.* "Workman with a water bag—during construction of Boulder Dam," 1934 (photograph, Ben Glaha, Bureau of Reclamation). Courtesy of the Boulder City Museum and Historical Association.

RIGHT: *Figure 6.07.* "Construction of Boulder Dam, Boulder City, Nevada—A typical Boulder Dam laborer—a young man, without a shirt, standing full-length, holding shovel," 1934 (photograph, Ben Glaha, Bureau of Reclamation). Courtesy of the US Library of Congress, Prints and Photographs Division, Washington, DC.

destroyer and life-giver—man needs the water to sustain his life while, at the same time, he builds the dam that will be used to control the river. In "Workman with a Water Bag" the expression of male sexuality and power is particularly evident in the vertical bar on the left side of the image, echoing Barnhurst's description of male virulence being evoked by phallic objects and forms. The straight line is a product of science and engineering, not of nature, while the foregrounded coded phallic imagery adds to the singular character of the photograph.

One of the most provocative images of Hoover Dam laborers is "Boulder Dam Laborer Holding a Shovel" (figure 6.07), an image that unmistakably conveys what Barthes describes as the "the *kairos* of desire," the right moment in which desire is a blissful eroticism.[51] This image also evokes a quality of "thereness" described by Barthes in *Camera Lucida*. To Barthes, when he (as referent: the photographed, the *necessarily* real thing that has been placed before the lens) knows he will be photographed, he cannot help but contrive pose after pose that he is inevitably unhappy with. He would rather be painted onto a "classic canvas" endowed with an intelligent expression by the artist than to be photographed. For him, subjects who know they are being photographed become "neither subject nor object but a subject who feels he is becoming an object."[52] However, in this image the photographer has managed to capture both the "thereness" of the individual and the *kairos* of his desire. In this image, the worker is, in fact, not even working but posing.

The man is young, suggesting innocence and virility. His body is an idealized form, sculpted and beautiful. Here we see his whole body (no part of it is cropped out of the frame), and yet his is the only body in the image; therefore, we cannot compare his size to another person, the dam, or the canyon. He is bare-chested and wears only work boots, hard hat, gloves, and close-fitting pants. The man's expression is neutral; he is not taken by surprise, nor is he photographed without an awareness of the photographer. His pose is relaxed and confident, leaning on one leg and using the shovel for balance. However, unlike in previous images, this laborer's gaze is not averted, nor is his face hidden. Like Manet's *Olympia* (1863), considered promiscuous because her gaze directly engages the spectator, this worker also looks intently at the camera.

Here we can also consider the laborer's accoutrements and the setting of the image, which also connote important cultural or social meanings.[53]

The tools that one carries are a symbol of one's economic and social status. Since ancient times men have been depicted in painting and sculpture holding tools: implements of war such as swords, spears, and guns; implements of agriculture such as hoes, rakes, and pitchforks; implements of labor such as work gloves, hard hats, and machinery; and so forth.[54] In this image we see the shirtless, muscled laborer holding a shovel, a low-skill tool in no way indicative of modernity or machine-age technology. The shovel, the hard hat, the work boots and gloves make it obvious that this person is not an engineer or a company executive. The setting is mostly obscured as the image is framed to show only the man and little contextual scenery other than the concrete and dirt below his feet. We do not see the dam, or the river, or machinery of any kind, but we do see a low smoothly shaped barrier behind him and in front of a sheer rock face. Little in the background indicates that this is a work site other than the writing etched into the wall in the background indicating a type of measurement. Here we must look to the caption, which tells us this is a "Boulder Dam laborer." It is only from the caption that we know it as one of the many images of the dam. It is only after reading the caption that we can place it in the context of its circulation.

Traditional depictions of men show them fully clothed, but since dress is central to the portrayal of people, clothing shown in any degree of undress is significant.[55] While the image of an unclothed female body always connotes sexual messages, the significance of erotic male nudity is expressed by a simultaneous state of dress and undress, what Anne Hollander describes as a "dialectic of clothes and body."[56] This particular image evokes a sexual tension that is not evident in other images of shirtless laborers, as the shovel in this image simply amplifies the sexual suggestion of the photograph. The shaft of the shovel extends from the ground past his groin up to his hand. The sexuality of wielding a shovel from the hip is apparent, but the thick, dark shadow cast by the shovel's handle emanating upward from his groin toward his torso, while producing a V that directs the viewer's eyes to the groin area, creates a multilayered phallic fantasy.

This image and others of muscled, shirtless laborers that suggest the traditionally feminine position in Western art imply a sense of homoeroticism that has not gone unnoticed by scholars. However, one check against homosexual readings of this type of imagery in the 1930s is the sheer pervasiveness of images in which shirtless, well-muscled workers appeared. Given the ubiquity of such images in New Deal art, it is difficult to consider them in terms of latent homosexuality.[57] Instead, some have proposed the term "homosocial," arguing that these images portray the intense male bonding experiences of hard labor. Social conventions of acceptable behavior and appearances simply would not have tolerated such open displays of deviance if the general population had interpreted them as homosexual. Thus, scholars caution against the mistake of imposing contemporary stereotypes of homosexuality and homoeroticism of excessively muscled men onto accepted social codes of the 1920s and 1930s.[58]

The Miniaturization of Human Labor

A February 1939 *Popular Mechanics* article entitled "Harnessing America's Wildest River" extends the sublimity/humanity dichotomy by intensifying the dominance of the technological sublime over both nature and people by twisting the Hoover Dam worker into something of a carnivalesque figure. According to the article, "so thoroughly has the Colorado been conquered that the *puny hands* of a few score men, backed by the *mighty strength of the dam*, now direct the almost limitless force of this 1,700 mile river and control the flow of billions of gallons of water almost as easily as you control the water at your kitchen faucet" (my emphasis). The article goes on to echo Burke's natural sublime by describing the Colorado River as "raging," "evil," "unruly," "a roaring torrent," "surly," a "renegade among rivers," "ungovernable," and "wasteful and destructive"—all descriptions that show it to be terrorizing, forceful, and powerful beyond imagination. However, the fact that this terrifying entity can be "tamed" and "thoroughly conquered" through engineering creates a sense of amazement—imagining that the terrifying power of the Colorado, a power that has menaced man for centuries, could now be harnessed so thoroughly that an individual person (most likely a female homemaker, no less) can control it from one's own kitchen sink.

But in this article, the human body is represented as not only miniature but also grotesque, a surreal inversion of the normally ordered state of affairs. Mary Russo comments that "images of the grotesque body are precisely those which are abjected from the bodily canons of classical aesthetics.... The classical [beautiful] body is ... closed, static, self-contained, symmetrical and sleek.... [T]he grotesque body is ... multiple and changing.... [I]t is identified with ... social transformation."[59] The grotesque body is also one of highlighted individual parts removed from the whole person, and in the *Popular Mechanics* article we see the human characterized by one part— "puny hands," which are presented as independent of a human being.

Parading of the grotesque often involves displaying an exaggerated element isolated from the whole.[60] The puny hands are a disfiguration and function metonymically as the group of laborers who work at the dam. The article includes a photograph in which we see a tour guide and several well-dressed visitors examining a switchboard in a control room at the dam. The tour guide displays the board proudly with a broad sweep of his hand, while the visitor's hands are clutched tightly by their sides or hidden from view. This anodyne room is the place from which man's technology controls nature through virtually no effort by the operator. The article confirms this by stating, "Yet little more effort is required to open a valve weighing several tons than is needed to turn the steering wheel of your car, because the water does most of the work. In fact, operating some of the valves and gates requires *no manual effort at all*. They work automatically" (my emphasis).

Hands themselves have been understood as an important part of giving voice and agency to people since classical times. Cicero mentions hand gestures in his discussion of *actio* in his *De Oratore*, and he returns to the use of hands again in *Orator*. According to Quintilian in his *Institutio Oratio*, without hands, one's ability for agency and discourse would be incapacitated since hands are almost as expressive as words. Although other portions of the body may help the speaker, to him, hands may almost be said to speak.[61] Thus, the "puny hands" of laborers become enfeebled and agentless, their usefulness overtaken by technology. Quintilian writes, "Do we not use them to demand, promise, summon, dismiss, threaten, supplicate, express aversion or fear, question or deny? Do we not employ them to indicate joy, sorrow, hesitation, confession, penitence, measure, quantity, number and time? Have they not power to excite and prohibit, to express approval, wonder or shame? Do they not take the place of adverbs and pronouns when we point at places and things? In fact, though the peoples and nations of the earth speak a multitude of tongues, they share in common the universal language of the hands."[62]

Quintilian goes on to write of the importance of gestures and motions of the hands in oration and offers detailed accounts of both the emotional effects best produced by hand manipulation and the mistakes orators make in using the wrong kinds of hand gestures. Hands, to him, are then

Figure 6.08. "Close-up of man working on construction of Hoover Dam," ca. 1931 (photograph, Bureau of Reclamation). Courtesy of the University of Nevada, Las Vegas Libraries, Special Collections. Elton and Madelaine Garrett Collection.

eminently important for a person's ability to speak, to demand, to show joy or sorrow; indeed, hands produce a universal language of their own. Consequently, a person without hands or with enfeebled hands, as is detailed in the *Popular Mechanics* article, is denied nearly all discursive agency. In figure 6.08 we see a visual metonym for the anonymous laborer, in which dirty and calloused hands are highlighted to document the process of construction, even as his face (and thus his identity) is excluded from the frame. Here, the important aspects of labor (hands) and construction (rocks, tools, and wooden forms) are what demand the photographer's lens.

This visual and verbal imagery reinforces the narrative of the dam as a state-sanctioned exemplification of the differences between classes and the accumulation of specific knowledge by certain groups at the exclusion of others (which I discuss in the next section), in which the giganticism of the dam and the sublimity of its technology dominate the puny human. Visitors to the construction site often described their experience in just such terms. Dorothy Childs Hogner explained her experience of seeing laborers at the dam this way: "In our sputtering motorboat we were like some small water bug beside this highest dam. . . . Above us, on the crest by the intake towers, we could see others of our kind, little creatures dwarfed by the size and magnificence of what man had erected."[63] Likewise, D. L. Carmody, in an article in the September 7, 1931, *Las Vegas Nevada Evening Review-Journal,* wrote, "Down close to the river was quite a colony of tent houses, mess tents, machine shots, and men—always men. From the boat we could see them on either side of the river appearing no larger than ants, looking so tiny that it was with amazement that I realized what had already been accomplished."

The laborers working at the dam site were routinely described as "small" and "tiny." They were mere insects compared to the massiveness of the canyon and the dam. Bruce Bliven, in the first-person narrative travelogue article "The American Dnieperstroy" for the *New Republic* in December 1935, also wrote similarly of his Hoover Dam encounter.[64] "Even from our height, a little below the top of the dam itself, the workers moving about amid their rubbish heaps seem *incredibly small and weak;* the mind refuses to grasp the fact that these *tiny dolls* and others like them have actually turned aside the river, built that vast cliff of concrete, chiseled and blasted out all these roads, many of them cut into the very face of the cliff, bored the mighty tunnels through which the entire river is now racing silently, far below our feet in the bowels of the hill" (my emphasis). Lynn Rogers, visiting the dam for the Outdoor Section of the *Los Angeles Times* in 1934, described his impression of the work site in similar terms: "Looking down on the vast structure as it rises to completion . . . the workmen engaged in power-house construction at the downstream base of the dam were as *mere ants* in the general scene" (my emphasis).[65] Not surprisingly, workers were also likened to spiders, and the massive construction machinery to "mechanical giants."[66] In the January 1, 1934, edition of the *Los Angeles Times,* an article with the headline "Foot-by-Foot Gigantic Boulder Dam Steadily Grows" includes a caption to a photograph depicting the base of the slowly rising dam that reads "TINY FIGURES OF MEN work on the top of the structure emphasize the immensity of the dam," which again echoes other juxtapositions in which human figures are depicted as tiny in comparison to the gigantic dam.

Photographs also perpetuated this metaphor of miniature humans. Conjuring notions of six-inch-tall Lilliputians diligently working to capture Lemuel Gulliver as he slept by tying his arms, legs, and hair to the ground with pieces of thread, these images show moments in which the miniature subjects, laborers nicknamed "high-scalers"—called such partly because they earned a slightly higher wage (75 cents per hour) than other laboring jobs—are set against the gigantic natural backdrop of the canyon walls (figure 6.09). The high-scalers' task was to rappel down the canyon walls on ropes to remove loose rock and to drill dynamite "powder holes" with jackhammers. The web of ropes holds the laborers suspended in midair as they place dynamite and excavate rock in preparation of the canyon to receive the dam. These photographs show the high-scalers sitting on their "bosun's chairs" dangling from ropes, clearing obstructions with dynamite and jackhammers for the eventual joining of dam ends to the canyon. The typical composition of these photographs emphasizes both the height at which they are working and their minuscule size compared to the massiveness of the canyon wall as indistinguishable human forms affixed to the rocks.

Echoing Barthes, Susan Sontag observes of the photograph that "it is always the image that someone chose," further adding that "the camera's rendering of reality must always hide more than it discloses."[67] The central fact of photography, unlike painting, is that the image is not necessarily conceived, but selected. The physical apparatus of the camera forces an act of isolating from the surrounding context an image that creates a new relationship that did not exist before. Sontag writes that "photographed images do not seem to be statements about the world so much as pieces of it, miniatures of reality that anyone can acquire."[68] But the miniature is marked by an ideology of distance, of an "over-seeing" viewer in which there exists what Susan Stewart describes as a "transcendence of the upper class, the reduction of labor to the toylike, and the reification of inferiority."[69] In Hoover Dam imagery the "acquisition" by the camera that Sontag describes contributes to a narrative of the oppression of the lower class and frames the human figure as miniature in relation to nature and, as we will soon see, to technology.

In these types of images we see the tiny laborers working to conquer the seemingly unconquerable, and although we know that the men are busily transforming the wall of rock for the new function it will serve, their individual activities are obscured and consumed by the landscape. The workers are enveloped by the gigantic; they are surrounded and enclosed by both nature and technology. While the scale of the background and the perspective of the camera fuse man and rock, the individuals become insignificant. The only importance these bodies hold is in the aggregate, performing their dangerous labor en masse. In these kinds of images, they are captured by a

Figure 6.09. "High-scalers—Hoover Dam site," ca. 1932 (postcard, Bureau of Reclamation). Courtesy of the University of Nevada, Las Vegas Libraries, Special Collections. Manis Collection.

camera and catalogued by a chronicler not as individuals, but as novel curiosity owing to their minuscule appearance set against the canyon.

Another characteristic of these and other similar photographs is the distance between the photographer and the photographed. The enormity of the dam structure and surrounding landscape often required photographers to move far back in order to frame their shots. This further heightened the disparity between the gigantic dam and the tiny humans. In *The Poetics of Space*, Gaston Bachelard comments that distance itself "creates miniatures at all points on the horizon." He likens the effect of distance on writers having "belfry daydreams" to such hackneyed phrases as "the men look 'the size of flies' and move about irrationally 'like ants.'" He continues: "From the top of his tower, a philosopher of domination sees the universe in miniature. Everything is small because he is so high. And since he is high, he is great, the height of his station is proof of his greatness."[70]

Figure 6.10. "First 8 cubic yard batch of concrete being placed in Boulder Dam proper," June 6, 1933 (photograph, W. F. West, Bureau of Reclamation). Courtesy of the Boulder City Museum and Historical Association.

The photographs of Hoover Dam laborers are also typically candid shots of men scaling the canyon walls, pushing concrete, or operating machinery, most of whom are unaware of the photographer's presence. Candid photographs must catch people as objects, oblivious to the photographer, acting in a "natural" (unself-conscious) manner—that is, having no control over their depiction. The panoptical nature of the work site itself facilitated such candid shots. Like prisoners in Bentham's Panopticon, these men did not know when, how, or from where they were being viewed or their picture was being taken.[71] In figure 6.10 we see an image in which a group of "puddlers" are spreading the first concrete delivered into the forms by buckets, eight cubic yards at a time. After each bucket of concrete was delivered, puddlers would walk around inside the block or use pneumatic vibrating prods to pack down the concrete to make sure there were no air bubbles. The photo is taken from far above, and again we see the referents as miniature and unacknowledging of the camera. The image is evocative of a prison yard, with the angular wooden form surrounding the workers and the vertical wooden planks rising around the sides like prison bars. One of the by-products of this type of work site is, as Foucault suggests, *imposing* efficiency. The panoptic nature of the canyon work site fits with Foucault's description of a panoptic structure, which is "polyvalent in its applications; it serves to reform prisoners, but also to treat patients, to instruct schoolchildren, to confine the insane, to supervise workers, to put beggars and idlers to work. . . . Whenever one is dealing with a multiplicity of individuals on whom a task or a particular form of behavior must be imposed, the panoptic schema may be used."[72]

Engineer as Hero-Ideal

These images also symbolize the diminishing role of manual laborer in both the economy and the public consciousness during the 1920s and 1930s. The Industrial Revolution initiated a profusion of manufacturing techniques and automation that replaced skilled workers with machines and factory assembly lines. This is especially important when we consider the principal material used in the construction of Hoover Dam—concrete. Developments in concrete technology in the early 1900s allowed a more mechanized form of production with greater speed and efficiency. Machines made the slump uniform in every batch and allowed the delivery of many yards of concrete at one time while doing away with the need for wheelbarrows or other labor-intensive means of delivery. The mass production of everything from the rock aggregate to the concrete forms meant that the need for specialized knowledge was shifting decisively away from the laborer to the engineer. No longer were years of apprenticeship required to learn just the right mixture of sand, lime, and water, or how to get bricks to line up just right. All that was required was a body, a "puddler," to move the mixture around in the form to remove the air pockets—everything else was prefabricated and pre-designed by engineers.

As the transition from manual labor to machine labor gained momentum, bodily strength was becoming increasingly irrelevant. Instead of the brawny laborer, it was the engineer and his brain, or the business tycoon and his machines, that were becoming the catalysts of construction and economic growth. Some of the most important developments in this regard were the advancements in materials science and engineering in the early 1900s, particularly the improvement of rigorous standards and testing methods, gauging instruments, and specifications that were developed in laboratory situations but were able to be distributed and replicated in the field as quality control. There was a two-way street of science offering increased productivity to industry, but industry also providing unprecedented opportunities in scientific and technical occupations.[73] The university systems were the training grounds for many engineers, who would then carry their testing methods and procedures out into the field, where a college-educated quality control tester could make up to eight times the salary of a laborer. However, the dissemination of their specialized technical knowledge did not pass down to the non-college-educated workforce. Engineers were trained into a system of self-perpetuation that reflected highly on their college programs but did not circulate new knowledge to those working in the field.

A May 1934 *Travel* magazine article titled "Exploring America's Mightiest Dam: How Heroism and Skill Build Boulder Dam" describes a scene in which the author, Andrew R. Boone, shadows engineers working at the dam site. He writes, "Men in long boots labored in fresh concrete, carpenters hastened the building of forms, and more engineers took measurements." In the article, the author repeatedly describes the engineers as "efficient" and "clever" but does not extend such descriptions to the laborers. He suggests that the engineer's main job, at least on this particular visit, was to explain the intricacies of the dam and take measurements of various sorts, most notably temperature measurements to gauge the cooling of the concrete.

These quality control procedures and engineering specifications were part of the wider leap forward in American building and manufacturing processes taking place at that time. The Hoover Dam project came at a point in which specific knowledge and means of production were progressively being transferred from skilled craftsman to college-educated engineers.

Early 1900s American popular literature also perpetuated the notion of the engineer as hero, as well as larger notions of an American ideal of a dynamic and technologically based society. The popularity of the hero-ideal was plainly evident in fiction and nonfiction literature throughout the early 1900s, as American culture embraced the hero-engineer as the quintessential civilizer of society. But although popular literature fostered romanticized hero worship of engineering, technology, and efficiency, labor unions and others objected to what they saw as the dehumanizing force of large-scale subjugation of workers. Nevertheless, the technological triumphs of the early 1900s contributed to the public's growing admiration for engineers. The engineer was seen as inherently ethical and impervious to corruption, and because this belief was so pervasive, most every occupation would appropriate the title of "engineer." Celia Tichi describes the public sentiment of the engineer as "a figure of action and enterprise, of personal restraint, and especially of unassailable ethics [which] made the engineer the perfect occupational namesake, especially for those involved in trade and commerce since, as engineers, they could use semantics to allay customer suspicions about the quality of materials in the margin of profit."[74]

Although working on the dam was not for the timid, and the laborers were often admired for their bravery, the most invigorated praise was reserved for its engineering accomplishments. The Hoover Dam project, which would eventually become a key catalyst in the economic development of the southwestern United States, was uniquely suited to showcase the engineer as the new orderer of nature. The engineer transformed a vast and inimical desert landscape, and a Colorado River mythologized as "rampaging," "evil," and a "menace," into a utilitarian amalgamation of natural and unnatural composites, thus reshaping the recklessness, disorder, and chaos of nature into an icon of efficiency and modernist machine aesthetic.

In early 1900s America, the engineer was the agent of scientific management and technological advancement, while the engineer's practices were based on an ethic of efficiency inspired by Taylorism and Fordism. Fredrick Taylor's *The Principles of Scientific Management* (1911) exemplified the ethos of productivity, in which he appeals to Theodore Roosevelt's call for "national efficiency," and established efficiency as the guiding principle of a modernizing society. President Herbert Hoover, too (an engineer by training), was frequently referred to as the "Engineer-President." In a June 17, 1933, article on Hoover Dam, journalist Chapin Hall of the *Las Vegas Evening Review-Journal* wrote, "Engraved on a modest little monument, is the sentiment proclaimed by the Engineer-President Herbert Hoover. . . . 'Here man builds his vision into rock that future generation may be blessed.'" The engineer had emerged as the quintessential symbol of power and control, able to fashion concrete and steel in the service of his own vision and to make that vision available for all to see. The engineer was the man in charge: controlling, shaping, and physically manifesting his designs through the manipulation of nature through technology.[75]

Tichi describes the early twentieth-century engineer as "no longer a craftsman fashioning unique objects from raw materials, but a designer committed to the functional, efficient arrangement of prefabricated components into a total design."[76] The engineer was modern, visionary, scholarly, and efficient. The engineer at the turn of the century was the new hero figure, replacing the cowboy of the nineteenth century, yet displaying many of the same characteristics of the mythical rugged individualist conquering and taming the wild world he encountered. Moreover, in contrast to self-serving executives, bankers, and politicians, the engineer was dedicated to altruistic projects such as building canals, bridges, and dams for the good of all, not for the sake of profiteering.

Class-Reinforcing Imagery

The increasing separation between the classes of the hero-engineer and the traditional laborer is visually articulated in a series of photographs taken during the dam's construction in which we see deliberate arrangements of economic capital and social stratification that reinforce implicit class hierarchies. This arrangement is manifest in a series of commemorative photographs taken between 1931 and 1935 that show engineers and VIPs, while the others present milestone markers of concrete poured.

Thus far, we have seen photographs of dignitaries posing inside of penstock pipes, laborers toiling in the concrete or hanging from the side of the canyon walls, but no images with both groups present at the same time in the same photograph. Most commemorative photographs, for example, have typical composition in which engineers and government officials are arranged across the image with no laborers in the frame. These next few images, however, are some of the very few in which laborers are in the same frame with engineers, business executives, or government officials. In spite of this, none of these mixed-hierarchy photos include laborers as a focal point or even as engaging the camera; instead, laborers are presented as either deliberately anonymous (included in the frame but unidentifiable) or deliberately disengaged (included in the frame but busy working and disengaged from the camera while the others pose for the photograph).

In the first photograph, "Group of men including Frank Crowe at the dam site" (figure 6.11), we see seven men (Frank Crowe is in the forefront of the photo, hand in pocket). Six of the men stand in a line in front of several planks of wood, all wearing business suits, and three wearing hats. Each man stands relaxed and looks intently at the camera; three men have their hands behind their backs, while the other three have hands at their sides or raised in the act of smoking. The seventh man, however, is dressed differently. He is dressed in baggy work clothes and a disheveled hat. He is off in the background not looking at the camera, but instead stands with his back to the camera, gloved hands held high as if directing activity somewhere off in the distance. This could merely be a coincidence—dignitaries having their photograph taken while work continues in the background, and a laborer just happened to wander into the shot. But notice the suited man on the far left. He is leaning into the frame. His body is at an angle, full weight on his left leg. Why would the photographer not simply adjust the camera frame to the left to fully include the suited man and cut out the laborer in the background completely? This image shows activity in the background—a laborer directs action, a boom from a piece of machinery lurks in the shaded area, and a cable stretches past the worker's hand. Yet only the suited men are identifiable; only the suited men are allowed to gaze into the camera; the viewer is only allowed to gaze at the suited men. The laborer remains anonymous.

The visual and verbal interactions of the next image construct a narrative through the use of numerical captions and signage within the frame of photographs documenting the event: the two-millionth cubic yard of concrete used in the dam's construction. Here we again see well-dressed men standing in a row and directly engaging the camera. A few laborers stand off in the background but are too far back to be recognized and so remain safely anonymous. Since the photographer could have moved the subjects to

Figure 6.11. "Group of men including Frank Crowe at the dam site," ca. 1931 (photograph, Glenn A. Davis, Bureau of Reclamation). Courtesy of the University of Nevada, Las Vegas Libraries, Special Collections. Glen A. Davis Collection.

Figure 6.12. "Second millionth yard being placed in Boulder Dam," June 6, 1934 (photograph, Bureau of Reclamation). Courtesy of the University of Nevada, Las Vegas Libraries, Special Collections. W. A. Bechtel Collection.

include everyone in the photograph, we can assume that they are intentionally excluded (figure 6.12). In this and other similar images in which milestones were commemorated with a photograph, the technology is displayed prominently with, for example, the narrative signage hanging from an eight-cubic-yard concrete bucket. Here the engineers and dignitaries stand in a row on wooden boards laid across the concrete so as to not get their feet stuck or shoes overly soiled. The laborers do not engage the camera, as their heads are either turned away or are too far away to be in focus; thus, they remain anonymous while our attention is drawn to the technology that fills the center of the photograph.

There is, however, one commemorative photograph that seemingly violates these norms. "Day Crew No. 7 cableway Record Pour" (figure 6.13) shows laborers arranged in front of canyon rocks and cableway machinery. All of them are looking directly into the camera, many of them smiling. This photograph, nonetheless, was taken only to celebrate the workers'

Figure 6.13. "Record Pour—Day Crew on No. 7 Cableway, 277—8 C.Y. Buckets—8 Hour Shift—12-30-34, Boulder Dam," December 30, 1934 (photograph, Bureau of Reclamation). Courtesy of the University of Nevada, Las Vegas Libraries, Special Collections. W. A. Bechtel Collection.

record-breaking production: they poured 277 buckets of eight-cubic-yard concrete in one eight-hour shift. This image celebrates the efficiency ethos of Taylorism, Fordism, and scientific management practices, establishing the cause for commemorating the ever-increasing output, not the mere existence, of labor. In keeping with past norms, however, we do not see any suited engineers, government officials, or dignitaries in the frame with the laborers. Moreover, in contrast to other commemorative photos depicting engineers and dignitaries, we are also not provided a detailed list of the names of each person in the image. In fact, no one is named, and thus all of the laborers, though photographed, remain anonymous.

THE ALIGNMENT OF WORK and moral virtue has been a fundamental theme for Americans since the seventeenth century. In the late 1800s and early 1900s, rising discordance over what work and labor meant to Americans constrained the role of laborers to either economic producers or what Erika Doss calls "wage slaves."[77] Labor unions defined the collective struggle for better working conditions and strove for unionization that promoted equality and economic balance, while industrialists and business leaders pushed their own ideas of labor centering on consumerism, profitability, and laissez-faire governmental regulation. The image of the professional, white-collar industrialist clashed with the eroticized and objectified bodies of laborers in both American culture and the visual representations of that culture.

This clash reached new levels in the 1930s, an era of recurring scenes that reinforced stereotypes of white-collar and blue-collar workers. These scenes depicted symbolic images of "manly" workers, often as eroticized objects of sexual desire, which reveal the degree to which the visual culture of the 1930s embodied wider anxieties about the changing nature of work and the role of the American working class in a quickly modernizing society.[78] New Deal agencies worked to counter this sentiment by courting artistic iconography of labor and commissioning upbeat images of rugged workers in order to visualize a sense of community and collectivity for a depressed and anxious public.

During this time period, we have seen two competing iconic myths emerge: the hero-engineer and the idealized "manly" worker, both of which are expressly visualized in Hoover Dam imagery. However, the imagery of seminude workers in art and photographs of the 1920s and 1930s, and also expressed in Hoover Dam imagery, suggests an anxiety about not only the future of human labor but also the hierarchy of work at that time. A second representation in Hoover Dam imagery, one of the disenfranchised working poor, rarely made it into circulation, as government and business interests controlled the means of distribution and worked to sequester images that might have cast the dam project in a negative light.

Although many depictions of laborers show them as foils to technology and subservient to the hero-engineer, laborers were still widely celebrated and venerated for their contributions to the project. And though the laborers were a group entirely composed of men, and almost entirely white men, there were other people—women, Native Americans, and African Americans—who lived and worked at the dam site, too. Those groups did not receive accolades in plaques or statues, or New Deal-era artworks. Moreover, it was mostly images of dutiful laborers who embodied the aesthetic appeal of the "manly worker" that were celebrated in images controlled by the government and circulated by a complicit media.

7

Advertising and Tourism

When construction ended, the purpose of the visual and verbal depictions of the dam shifted quite dramatically. No longer celebrated for controlling flooding, providing drinking and irrigation water, or bringing economic windfalls to Southern California, and with its utilitarian functionality achieved and the heroic laborers and engineers long gone, the Bureau of Reclamation turned to reimagining and heavily publicizing the dam area as a tourist and vacation destination. This move toward tourism relies heavily on two themes: secular pilgrimages to experience the technological sublime firsthand, and a "playground" destination for camping, fishing, and boating opportunities, particularly for people living in Los Angeles. No longer the arid "cactus-covered waste" President Franklin D. Roosevelt describes in his dedication address, the vast space covered by the newly formed Lake Mead was reimagined as a place for family vacations, fishing, and water sports.

While the message of American imperial sovereignty over what was once desert but is now "reclaimed" would continue to spread in the months and years following the completion of the project, the verbal and visual discourse about the dam assumed a variety of new forms and purposes. As the situated documentary photography of the dam's construction gave way to a verbal and visual rhetoric that centered on a new aesthetic of tourism and recreation, the dam's by-product—the newly created Lake Mead—became heavily promoted as a vacation destination. Consequently, many of the sorts of images used to encourage tourism also seeped into the advertising campaigns of various commercial products. Beginning with business-sponsored newspaper outings to the dam site during the late construction and early post-construction periods, the dam became a commoditized advertising backdrop designed to draw on what Roland Barthes calls "connotative external codes" already circulating in the public consciousness about the dam. Although the dam was celebrated as a technological wonder, it was also a source of national pride, and advertisers sought to exploit those sentiments and invigorate consumerist behavior by linking the nation's pride in the dam with a drive to fulfill industrial imperatives.

Hoover Dam Imagery in Advertising

By way of their unyielding control of images and texts of the dam, the Bureau of Reclamation's publicity effort was clearly successful. The effect was such that the bureau not only instilled public confidence in the dam and its own organization, but also persuaded the public to see the dam as a symbol of American ingenuity, efficiency, power, and engineering acumen. Consequently, advertisers saw an opportunity to take advantage of the positive associations the bureau had developed with the public regarding the dam. Cigarette companies, alcoholic beverage makers, car manufacturers, paper millers, oil companies, steel manufacturers, airlines, travel companies, heavy industry, and many others published materials connecting their businesses and products visually and verbally to the project.

Trucks were an especially popular item for which Hoover Dam was used as part of a sales pitch. Alcoa Aluminum boasted that their new "modernized" truck beds were speeding industrial progress by being lighter and stronger than comparable steel ones, and that a truck with "an ALCOA aluminum body that dwarfs all others, is literally moving mountains at Boulder Dam." International Harvester Company, maker of International Trucks, writes in one of their advertisements, "Six Companies Inc., Builder of Greatest Engineering Project Since the Panama Canal, Place Large Order Exclusively with International Harvester." The advertisement includes images of the river at the bottom of the canyon just as construction on the diversion tunnels was beginning, as well as an image of an International Harvester dump truck, a rendering of the completed dam, and a small map showing the location of the dam on the Arizona-Nevada state line. The advertisement goes on to explain, "The full meaning of this decision—the extent of the honor paid to International performance and service—can be appreciated only when measured against the immensity of the project itself."

A brochure from the Zellerbach Paper Company titled *Progress*, promoting their Cumberland Gloss paper, also attempted to associate their product with the dam. The front of the folded brochure features a dump truck on each side, divided by a vertical panel in the middle that contains the word "Progress" in capital letters. When opened, the brochure features a short history of the Boulder Dam project and reveals the purpose of the booklet: "The Hoover Dam was selected for the initial broadside because of the gigantic engineering problem involved, the size of the project and the benefits that will be derived upon its completion." The theme of progress through technological advance is reinforced inside the brochure, where the massiveness of the project is expressed through a recitation of statistics about the Colorado River and Hoover Dam and a striking photograph of the excavation work for the cofferdam on the canyon, showcasing the paper's "photographic effect to the halftones" and "unusual strength and folding qualities."

Another company that used Hoover Dam as a backdrop to advertise its products was R. J. Reynolds Tobacco Co. Several advertisements for Camel brand cigarettes featured the dam. These advertisements, however, emphasize a different facet of the dam project—glamour. John Berger writes that publicity (and advertising) is a system with a single purpose: "It proposes to each of us that we transform ourselves, our lives, by buying something more. . . . Publicity persuades us of such transformation by showing us people who have apparently been transformed and are, as a result, enviable. The state of being envied is what constitutes glamour. And publicity is the process of manufacturing glamour. . . . Publicity is never a celebration of a pleasure-in-itself. Publicity is always about the future buyer."[1] As I mentioned earlier, the engineer—a powerful symbol of efficiency, technology, progress, and the promise of a better future—was highly regarded in the early 1900s. R. J. Reynolds drew on both the ethos of the engineer and the dam's construction to advertise its Camel brand cigarettes. The cigarette advertisements attempt to connect the glamour of engineering Hoover Dam to the thrill of smoking Camels.[2]

One such advertisement (figure 7.01) features testimonials and photographs of "outdoor people" such as rancher Charley Belden from Pitchfork, Wyoming, and Helene Bradshaw, "an enthusiastic horsewoman" who says,

Figure 7.01. R. J. Reynolds Tobacco Company advertisement, ca. 1935

"Every woman prefers a milder cigarette." The largest image—the only color image, and the one positioned closest to the Camel cigarette pack and logo in the bottom right—features Erwin Jones, "Boulder Dam engineer," reclining against a rock with a surveying tripod next to him and the intake towers of the dam behind him. Jones is quoted in the advertisement as saying, "It's been thrilling to have a part in the vast enterprise of building Boulder Dam. Plenty of strain, too. When I get tired, there's nothing like a Camel. Man, what a swell taste Camels have! Mild, cool, mellow! You can tell they are made from choice tobacco, because they don't get 'flat' or tiresome when you smoke a lot."

Another cigarette advertisement, this one a full-page comic strip, starts with the title "Building the Great Boulder Dam: A Gigantic Monument to American Energy." Underneath the title is the caption "A recent check-up shows that more Camels are smoked at Boulder Dam than any other cigarette. Camels give your energy a lift!" The rest of the comic shows the progress of the project from the first surveys, to blasting the diversion tunnels, to the work of the high-scalers, to the pouring of concrete, to the closing of the diversion tunnels and filling of the reservoir. All along the way Camel cigarettes are part of the story. As two men look down on the base of the dam, one reclines against a wall smoking a cigarette. As a group of men are hoisted down to the work site, one says, "The day shift's coming on, and boy am I tired!" to which another responds, "Me too—let's have a Camel." Next to a cross-section technical drawing of the dam, a man in a suit asks another man, "How do the men who built the dam stand on cigarettes?" to which the other replies, "Camels are the favorite." Finally, Erwin Jones appears again, this time identified as "Staff Engineer, Boulder Dam." He says, "Smoking Camels refreshes me in a very few minutes. Camels taste mild and mellow, you can tell they are made from choice tobacco."

Hoover Dam and Tourism

Since the eighteenth century, tourism has played a powerful role in the invention and development of America's cultural identity.[3] Following in the traditions of the Aesthetic Movement, the art of the Hudson School, and writings by Wordsworth, Thoreau, and Byron, nineteenth-century Americans' notions of landscape and culture became so closely aligned that the two ideas were nearly identical.[4] American secular practices of vacation and travel to witness examples of natural or technological sublimity in many ways paralleled pilgrimages to religious sites. Tourism provided a way in which America could be defined as a place in which Americans could take pride in the distinctive features of their landscape, one in which tourist attractions assumed the role of what John Sears has described as the "sacred places" of traditional societies.[5] Sacred places, sites that resonate with the most cherished values of that culture, have universally attracted some form of travel. Such travel often has a religious motive, as in pilgrimages to Mecca or Jerusalem, but can also be secular, such as in trips to national parks or monuments—though even in the latter case, religious veneration often plays a part. Patrick McGreevy has observed that in the nineteenth century Niagara Falls was one such sacred place. He writes, "What is peculiar is not only the things that accumulate in Niagara's landscape, but also the people who gather there—for example, the pilgrims who have journeyed to Niagara for religious reasons. The Catholic Church recognized these travelers in 1861 when it consecrated Niagara Falls a 'pilgrim shrine,' granting it the same official status as Jerusalem or Rome."[6] And just as religious pilgrims seek out Jerusalem or Mecca, Americans seek out their own holy sites, such as Niagara Falls, Yosemite National Park, and the Grand Canyon.[7] Tourist attractions in nineteenth-century America emerged as a reflection of a secular culture, and one progressively more oriented toward consumerism. Industrialization, urbanization, and an increasingly wealthy middle class enabled more than just the elite to travel, and this trend served to increase the need for travel destinations—that is, scared places.

A brochure titled "President Visits Hoover Dam" given to attendees of Franklin D. Roosevelt's dedication address mentions many of the familiar benefits of the dam project: water supply, silt control, power generation, flood control, and so forth. Although we have seen these claims many times before, one new section of the brochure is devoted to promoting the dam site as a recreational destination. It states, "A lake formed by Boulder Dam, unique in its beauty and depth of water, will attract tourists from the important Midwestern and Pacific coast cities who enjoy the pleasures of camping, boating, fishing, and bathing. Even now visitors to the project are coming in increasing numbers, a late report showing that tourists [sic] travel, which in 1933 numbered 132,646 visitors, has materially increased, and that during the first 6 months of the present year 184,983 visitors entered the reservation."

Early pre-construction arguments in favor of the project put forth by Clyde L. Seavey and others—arguments visualized in various booklets and newspapers, including the *Evening Herald* front-page image—seldom mention activities such as boating, fishing, swimming, and camping as potential benefits of the dam. But upon the dam's completion, the original arguments for the project, such as protection from floods and drinking water for Los Angeles, became afterthoughts as both the verbal and visual rhetoric of the dam shifted decisively toward the subjects of tourism and recreation.

One of the few pre-construction- or construction-phase instances in which the dam was mentioned as a tourist destination was in an article by

Secretary of the Interior Ray Lyman Wilbur in the January 15, 1932, "Progress Edition" of the *Las Vegas Evening Review-Journal* titled "Hoover Dam Will Create a Unique Tourist's Mecca." Wilbur writes that the dam and the region "will become one of intense national interest." "Not only will the dam, with the great body of water behind it create a situation such as man has never before seen," he continues, "but the area in Arizona and Nevada tributary to it will offer unique possibilities of development for public recreational use." In April of 1935, shortly before the dam was finished, the *Boulder City Journal* predicted four hundred thousand visitors that year. The front-page article is accompanied by a photograph of a man pulling a fish out of the impounded waters behind the dam. The caption claims to show the "first fish" caught in "Boulder Lake." No longer referring to it as the Colorado River, the article cites reports by the Bureau of Reclamation that the lake and dam had attracted 123,672 tourists in the first four months of the year, 266,436 in 1934 (up from 96,000 in 1932), and 615,000 visitors over the past four years.

A 1937 promotional brochure from the Boulder City Chamber of Commerce also touted the dam as not only the "World's Most Impressive Engineering Spectacle" but also "the Center of an Area of Unsurpassed Scenic and Historical Beauty." According to the brochure, the dam is "Man's Challenge to 'The World's Most Dangerous River.'" Another brochure invited visitors to "explore the virgin wilds made accessible by this spectacular lake," which is "situated in the most ideal climate in America." The brochure suggests that visitors see the "Caves and Old Indian Habitations" and remarks that "Lake Mead is destined to become one of the outstanding beauty spots of North America. . . . [I]t makes accessible by water . . . scenes of grandeur heretofore unglimpsed except by small parties of intrepid explorers."

Early promotion of the dam as a tourist attraction (such as I have just outlined) prefigures a later expansion into advertising. These two practices—tourism promotion and advertising—are particularly instructive of how images of the dam were used to shape Americans' sense of themselves in relation to the project. Additionally, the regional and national prerogatives of commerce and tourism played a significant role in shaping the types of images that were circulated to the public in the dam's post-construction era. Tourism and advertising images both reflected the economic climate of the period and helped shape American cultural values and identity.

Hoover Dam Tourism and the Great Depression

For mass tourism to exist it requires people with the money and time to travel, an adequate and affordable means of transportation, and circumstances of relative safety and comfort at the destinations. But it also requires a corpus of imagery and description of those places, what Sears calls "a mythology of unusual things to see" that invigorates people's imaginations and encourages them to actually go there.[8] The Great Depression era in America was a time in which money and leisure were scarce; however, several factors were instrumental in positioning the dam site as a tourist destination, even during the woeful economic situation of the 1930s.

As we have seen, mythology about the dam had been building starting in the early 1920s. Descriptions of a project so huge that it was beyond words accompanied by images that attempted to visualize the massiveness of the dam captured the imaginations of Americans and incited people to witness the project themselves. Additionally, the automobile was becoming increasingly available to middle-class Americans as a form of relatively inexpensive transportation. Finally, tourism and vacationing, particularly to state and national parks, was being endorsed as a low-cost relaxation and entertainment activity for Depression-era Americans.[9]

Sociologist John Urry, in developing his notion of the "tourist gaze," suggests that tourism is part of a socially constructed system to engage different environments with interest and curiosity. Urry argues, however, that "there is no single tourist gaze . . . there is no universal experience that is true for all tourists for all times. Rather the tourist gaze in any historical period is constructed in relationship to its opposite, to non-tourist forms of social experience and consciousness."[10] The tourist gaze of 1930s America was constructed in relation to the trauma of the Great Depression. By 1933 the stock market's value was barely 20 percent of its 1929 peak. As factories and businesses shut down, millions of workers were laid off. As banks failed by the hundreds, many families lost their life savings. Great migrations of families and individuals crisscrossed the country in search of work. Given this economic plight, one might think that vacations and tourism were practically nonexistent. However, despite the severity of the economic situation, vacations continued to be a popular American pastime. Newspapers and magazines continued to offer plenty of advice for families, much of which centered on not letting the Depression render them "vacationless."[11]

Publications such as the *Ladies Home Journal, Popular Science Monthly, Women's Home Companion, Good Housekeeping, Fortune, Harper's, Nature Magazine,* the *New York Times,* and *New York Times Magazine* all promoted the value and necessity of vacationing. Some argued, in fact, that because of the economic problems, vacations had become *more* important, not less. Many suggested that families take "stay at home" vacations to local parks, swimming pools and recreation facilities, golf courses, and even backyard barbecues.

Even first lady Eleanor Roosevelt wrote of the benefits of vacationing. In the August 1934 edition of *Women's Home Companion* she argued that vacations provided an opportunity for needed rest and rejuvenation. She wrote of one family's vacation to a nearby state park, where they rented tents by the lake, bought groceries from a local farm "at a very modest price," went fishing and hiking, laid in the sun, and read books—all inexpensive ways to find relaxation and enjoyment.

Tourism, Automobiles, and Advertising

For many years, destinations such as national parks were restricted to railroad-only access. However, after World War I, the growth of automobile ownership among the middle class made railroads virtually irrelevant as far as tourism was concerned. Frank Brunner wrote in *Outlook* magazine in 1924, "The automobile has revolutionized the average American's vacation. It has brought about a renaissance of the outdoors and it has firmly planted a brand-new outdoor sport . . . Autocamping."[12] As the automobile solidified its position as the travel vehicle of choice for Americans, destinations devoid of railroad access, previously unreachable except for the most privileged adventurer, became accessible and, in many cases, very popular.[13]

Numerous Depression-era articles described ways in which families could take inexpensive vacations; and many suggested the family car to be the ideal vacation vehicle. With roads and highways better than ever, and with cars becoming bigger, more reliable, and more comfortable, vacationing was just a matter of filling up with cheap gasoline, piling the family into the car, and driving away.[14] Some of the most popular destinations were state and national parks. One magazine reported that families could stay in Yosemite National Park for $1.50 per day. Others wrote of small hotels that dotted the highways throughout the West and South, most of which could be had for as little as $1.00 per night, including breakfast.[15]

With magazines, periodicals, and celebrities such as Eleanor Roosevelt recommending vacation travel by car to Depression-era families, Hoover Dam was one recipient of especially ample press coverage as a drivable getaway destination, particularly by the Los Angeles Times. The Times ran numerous full-page articles sponsored by car companies featuring images and narratives about the cars and journalists' accounts of road trips to the dam. The articles were for all intents disguised advertisements. In the Sunday, June 18, 1933, edition of the Times, for example, the Motoring and Outdoors Section featured a collage of photographs and hand-drawn illustrations by journalists who visited the "gargantuan project [from Los Angeles] in seven hours over splendid roads" (figure 7.02). In the article, journalist and artist Charles Owens remarks on the "comfort" in which the party traveled to Las Vegas and on to Hoover Dam. On the page are photographs of the car superimposed on illustrations of desert vistas and a map of the link between the Arrowhead and Old National trails, which would eventually pass over the top of the dam, forming a "circle tour for Los Angeles folk that will carry them from the green of the coast to the gray of the desert."

Another Los Angeles Times Outdoor Section in March of 1936 features "the 1936 Hudson eight sedan from Earle C. Anthony, Inc." Similar to the other Times articles, this one highlights the dam as a "Thrilling Sight" and features several photographs of the dam and car together, along with a map of a travel route. One photograph shows the dam with the Arizona side valves open in the background, another the highway across the top of the dam, while another shows the car parked at the boat launch for trips on the lake. What we might think of as more traditional advertisements also showcased cars in front of the dam. Studebaker ran an advertisement showing its 1940 Studebaker Commander parked in front of the dam capped with the slogan "This Is the Year to Visit Your Own America's Wonderlands" (figure 7.03). The advertisement copy argues that since Americans "can't travel abroad," then this is the year to get out and see America. The advertisement suggests that although this $965 car is a "spacious living room on wheels," it is also very affordable: it "runs hundreds of miles on a single filling of gasoline," is "thrifty," and "inexpensively stands up under driving that would make less carefully constructed cars cry quits."

In the image the Studebaker is parked in front of Hoover Dam, while several people stand, walk, or sit at various distances between it and the river. And although the huge dam towers over the car and commands the scene, not one of the five people looks at it; instead, they look at the car. One woman

Figure 7.02. "Comfort Marks Trip to Hoover Dam Site," *Los Angeles Times,* June 18, 1933. Courtesy of the Boulder City Museum and Historical Association. Copyright © 1933. *Los Angeles Times.* Reprinted with Permission.

Figure 7.03. Studebaker advertisement, 1940

in the background on the far left is turned toward the car with her back to the dam, while a man on the far right aims his camera at the car instead of the dam. Although the dam had been widely publicized as a world wonder and a monumental feat of modern engineering, the man chooses instead to take a snapshot of the vehicle.

Another advertisement, this one for the 1954 Chieftain Pontiac, also shows a car in front of the dam. The ad copy again highlights affordability. The low price is "a conspicuous Pontiac virtue," yet the car is "splendid enough inside to win the hearts of the most style minded." A stylized drawing in the top left is a close-up depiction of a woman in the driver's seat (foot on the brake instead of the gas) accompanied by a caption describing optional amenities. The larger image in the center of the page presents a full view of the car with the interior clearly visible, and this time with a man in the driver's seat, a woman in the front passenger's seat, and a child in the back seat. The child is turned around looking at the dam, while the man and woman (husband and wife presumably) have broad smiles on their faces, beaming with "pride in [their] Pontiac." Just as in the Studebaker ad, this one shows the adults ignoring the dam. In the Pontiac ad, the man beams while driving the car. The woman looks at him, not at the dam. The dam is too enticing only for the child. In each case, the adults (the consumers buying the cars) are so enraptured by the automobile that they do not even notice the dam.

What we are seeing here is a subtle change in advertising strategy over time. In the *Los Angeles Times* articles from the 1930s, the sponsor's car is mentioned, but the dam is the center of the story. Although the car's comfort, economy, and performance are explicitly mentioned in the text, the dam and surrounding areas are referenced repeatedly. Normally, in these images the reader is not enjoined to desire anything in particular. In the 1933 *Times* article, for example, there are only illustrations and vague photographs of roads and a water embankment, but the images and narrative are still predominantly about the dam. The 1936 *Times* article, however, is more visually impressive and more overtly persuasive in intent. The largest and topmost image on the page shows a man standing alongside his car in front of the completed dam. The scene looks peaceful: no massive construction site, no dust storms, only water gushing forth from the side of the canyon through the penstock valve. As opposed to the images in the previous articles, a newspaper reader in Los Angeles could envision themselves in the place of this person, calmly taking in the scenery and the dramatic panorama of the dam. The other images accompanying this article show towering cliffs reflecting off a calm waterway and cars cruising unobstructed across the completed dam.

Although the differences in presentation are subtle, we can observe in the more contemporary advertisements of the 1940 Studebaker and the 1954 Pontiac some clear divergences from the *Times* articles in how advertisers attempted to create structures of meaning. The buyer is meant to be envious, imagining their future ideal self transformed by the product and thus becoming the envy of others. The 1940 Studebaker advertisement clearly adheres to this principle more closely than previous *Times* advertisement articles. In it we see the dam, but we also see the car framed prominently at the bottom of the page. The people in the image are clean, well dressed, and confident—nonchalant even. The sight of the dam does not overwhelm them. Some of the people are shown casually strolling along, oblivious to the massive concrete structure. Some actively ignore the dam and instead pay full attention to the car. This image communicates to the spectator-buyer that they would be transformed by owning this car: if this car can outshine the greatest engineering project ever undertaken in the United States, then the spectator-buyer who owns this car would certainly be the object of others' envy.

In the *Times* travel articles we encounter a version of what we call today "product placement." A trip to the dam is the focus of the article, while the means of transportation (the sponsor of the article) receives a mention within the context of that story. The main "product," though, is still the dam itself. However, these and other similar advertisements would occasionally have entire sections or paragraphs devoted to describing the dam's construction or the history of the dam project, but that would fail to be integrated into a cohesive narrative that includes the product. Additionally, many of these advertisements require reading long blocks of text to understand their meaning. Conversely, instead of constructing a verbal narrative about products and their associated *activities*, the message in the Studebaker advertisement visually translates statements about the product into statements about consumers and human *relationships*.

According to noted advertising scholar Gillian Dyer, "Advertisements should be seen as structures that function by transforming an object into something which is given meaning in terms of people. The meaning of one *thing* is transferred or made interchangeable with another *quality*, whose value attaches itself to the *product*."[16] In the Studebaker advertisement, the image suggests that instead of making an arduous, albeit adventurous, trip to a construction site, by purchasing the car the spectator-buyer would instead enter into a relationship in which they would be envied by their fellow consumers. But because the caption still states that the photograph is of Boulder Dam, it leaves room for additional rhetorical development. Consequently, it is not until the 1954 Pontiac advertisement that we see the full evolution of advertising with regard to the dam.

Barthes argues that to create meaning, advertisements draw on *connotative* external codes—images, concepts, myths, and so forth—already circulating in society. This connotative level of analysis is what Barthes calls the "rhetoric" of an image. To associate a product with the dam was also to associate it with the dam's cultural mythologies, principally those related to advanced technologies and triumphant engineering. There is also a textual element to the rhetoric of these images. The *Times* travel editorials and the Studebaker advertisement mention the dam, but Pontiac, though showing an image of the dam, makes no mention of it in the text.

Advertisements do not just sell particular properties of consumer goods; they also seek to make those properties mean something to us, to create emotional attachments to the brand or product that are both involuntary and compelling. Advertisements provide us with a structure in which we are interchangeable with the goods we purchase; they are, in the words of Judith Williamson, "selling us ourselves."[17] We can see this in both the Studebaker and the Pontiac advertisements. Studebaker is selling envy; Pontiac is also selling envy, but differently. By using images of the family traveling past Hoover Dam with the child looking out the back window, but with no mention of the dam anywhere in the text, the advertisement highlights all of the connotative values of the dam without the burden of the denotative. The image of the dam encapsulates what Barthes describes as "the language of the image," which is "not merely the entirety of the utterances emitted . . . it is also the entirety of the utterances received; such language must include the 'surprises' of meaning."[18]

We can extend this insight to any number of advertisements that featured images of the dam. Some are quite conventional, others downright strange. As I mentioned previously, Hoover Dam was a favored backdrop of advertisements for trucks and machinery during the construction phase of the project, and that trend continued into the post-construction phase.

International Harvester, for example, against a backdrop of the dam with the jet valves open, proudly states that their trucks "handled 80 percent of the heavy hauling at Boulder Dam." The advertisement even features a commemorative plaque from Six Companies Inc. as a show of their "appreciation of service achievement to International Harvester in recognition of its co-operative contribution to the construction of Boulder Dam." Shell Industrial Lubricants uses a curious image of a draft horse pulling a thirty-three-thousand-pound weight through a series of pulleys inside a translucent cube suspended above the river. The text, in this case, elucidates the image's meaning: "Defined as a theoretical rate of work, one horsepower equals lifting 33,000 pounds to a height of one foot . . . in a minute" (ellipses in original). The advertisement goes on to explain that machines need high-quality industrial lubricant to perform efficiently, and that Shell should be industry's lubricant of choice.

Perhaps one of the strangest advertisements featuring the dam is for Calvert Reserve whiskey (figure 7.04). The caption placed below an image of the dam reads "World's biggest power plant—Hoover (Boulder) Dam *Challenges Comparison* with any work of man. Calvert's good taste *Challenges Comparison* with any whiskey!" (emphasis in original). I have already discussed many attempts to account for the dam's greatness by means of verbal, statistical, and especially visual comparisons; however, the presentation in the Calvert Reserve advertisement is nothing short of bizarre. In the ad, the dam sits in the background (valves closed this time), but it is somewhat off-center. The canyon wall on the right is actually the largest feature of the image. Above the dam the slogan *"Challenges Comparison!"* in yellow italicized text bookends a floating spaceship-like tray holding a Collins glass with ice on one side, a martini glass on the other, and a bottle of Calvert Reserve whiskey in the center.

As I mentioned previously, the Bureau of Reclamation worked diligently to create an aura of technological mastery, efficiency, and engineering prowess with regard to the dam project. Those notions, articulated through carefully edited representations and placed in wide public circulation, framed the public's interpretation of the project. Furthermore, the initial verbal arguments in favor of the dam, as well as the promissory and utopian visualizations of a post-dam Southwest, formed a mythology about the dam in the collective cultural consciousness of Americans. The dam project, artfully presented to the public through thousands of images over the course of

Figure 7.04. Calvert Reserve Whiskey advertisement, 1951

several years, provided advertisers a powerful symbol rich in connotations to which they could attach their products. For some advertisers, even seemingly odd associations with the dam were enough, as evidenced the Calvert Reserve whiskey ad, in which the connotative ideas about the dam make for a visual paring with the product.

Hoover Dam, of course, was built in large part for economic reasons (agricultural expansion of the Imperial Valley, population growth of Southern California, and so forth); thus, increasing goods and services for the Southwest was always a consequence of the project. However, the commodification of the dam's *imagery* through advertisements assigns an economic value to something that was not previously considered in economic terms. The dam that appears in the background of an automobile ad, not to celebrate the structure but to *ignore* it in favor of a product, both exploits the mythology surrounding the dam and works to alter or challenge it.

Tourism and Water Recreation

An article in the 1955 May–October edition of *Nevada Highways and Parks* describes the Colorado River in familiar terms: "Floods . . . brought death and destruction. . . . The river was too muddy for fishing; too swift for pleasure boating. . . . The stream, in large measure, was useless and dangerous. Its waters wasted into the Gulf of California for ages, with its power potential undeveloped." The dam, of course, addressed all of these issues; however, the article goes on to describe an additional, though unintended, outcome of the dam project: the "recreational centers, boat docks, tourist camps, lodges, and various other types of resorts . . . to accommodate any and all visitors who . . . recognize the value of this great body of water for water sports and recreational purposes." Numerous post-construction Hoover Dam tourism articles echo this sentiment by including images of families, vacationers, and outdoorsmen enjoying assorted water sports at what the Bureau of Reclamation was now calling the "Hoover Dam and Lake Mead Recreation Area."

One common metaphor describes the recreational area as a watery "playground." An article by Weldon F. Held in the July 1941 edition of *Arizona Highways* magazine titled "The Miracle of Lake Mead: Arizona's Magic Water Highway" refers to Lake Mead as an "Arizona playground." The February 1950 issue of *Desert Magazine* includes an article by Gene Segerblom titled "Desert Playground" and presents a half-page image of children swimming and splashing around in the water as if in a giant pool while others stand on or jump off of a floating platform. In the April 1950 edition of *Arizona Highways*, Joyce Rockwood Muench writes of "a bountiful national playground in the desert." Both the March 1956 and May 1964 issues of *Arizona Highways* also refer to the dam site as a "playground." Tourism at the dam site is the main subject of the 1956 issue, which includes several photographs of vacationers boating, swimming, and fishing on Lake Mead. The editor's introductory article, "Along the Colorado," states, "This issue we're going fishing, boating, swimming and we're going to have fun in the sun. . . . Here is America's newest playground, one of the biggest. . . . To the men and women of these services who are doing so much to develop this playground for us, are these pages respectfully dedicated."

A different article in the same 1956 edition, again by Gene Segerblom, states, "Up in the northernmost tip of Arizona there's one of the largest playgrounds in the nation." The notion of the dam site as "America's newest playground" is repeated in the 1964 issue, in which the introductory article comments that, although originally "intended for power and reclamation . . . even the most optimistic could not have dreamed of what popular playgrounds they were to become."

Two other activities were showcased in images promoting the dam site as a leisure destination: boating and fishing. The Outdoor Section in the December 3, 1933, edition of the *Los Angeles Times* features a large illustration (again by Charles Owens) depicting the "Times-Automobile Club party shooting 'flour sack' rapids in lower Granite Gorge of the Grand Canyon . . . [the point of the] extreme headwaters of the future lake created by Hoover Dam." The illustration shows several men in an "Airplane Seasled" powerboat speeding across choppy waves, with the canyon walls rising high in the background. Although the caption to the illustration says "this area, along the great Colorado River above the Hoover Dam site is one of the most scenic in the West," and the article reports that the "majestic walls of Granite Gorge" provide "vivid, colorful, thrilling views . . . of tremendous rock formations," the dramatic backdrop created by the canyon walls shown in the illustration would soon be under water, an irony that seems to have eluded the publishers.[19]

Images of people boating on the lake frequently appeared in travel magazines as well. For example, the August 1945 edition of *Nevada Magazine* features a full two-page color spread of an outboard motorboat with two men in the front and three women in the back speeding toward the dam and intake towers, which are soaring high above the slowly filling lake. The April 1948 *Nevada Highways and Parks* magazine issue on Boulder Dam includes several pages of photographs of powerboats, sailboats, fishing boats, boat docks, fishing harbors, a cruise launch, and even an image of a man "Fishing in the Modern Way"—from a seaplane on the water. The April 1950 edition of *Arizona Highways* features images of boat docks and motorboats on the lake, while other pages show children swimming. The May–October 1955 edition of *Nevada Highways and Parks* includes several images of vacationers boating, sailing, and waterskiing.

Fishing is another repeatedly visualized activity in dam tourism publicity photos. I noted previously the April 24, 1935, *Boulder City Journal* front-page article featuring a photograph of a man reeling in the "first ever" fish out of Lake Mead. From that point forward, imagery of people (typically men, though not always) smiling as they hold up their catch from the day's fishing excursion became a prominent visual trope for tourism promotion. One such image is from the February 1950 edition of *Desert Magazine*. In it, three men are shown holding up several large fish. Sunlight shimmers off of the fish, suggesting they are freshly caught; the men flash broad smiles, obviously pleased with their abundant catch. The April 1948 *Nevada Highways* magazine includes several images of boating and fishing. One image shows several shirtless "Hollywood Radio and Moving Picture Stars" wielding fishing rods on the back of a boat, while another image below depicts a row of six fish caught in the lake.

Women and children, too, were occasionally depicted fishing on the lake. The May–October 1948 edition of *Nevada Highways and Parks* includes the article "Lake Mead: Pitched a Fisherman's Paradise into the Lap of a Parched Desert," accompanied by images of men, women, children, and whole families angling in the lake's waters. One image shows four people (two women and two men) on a small pleasure craft; two of them are fishing—one woman and one man—while the others look on. On the next page a colorized image shows a young boy sitting in between two women at the back of a motorboat, all smiling broadly while holding up their ample haul of fish.

A nearly full-page image in the article "Where the Fishing Is Good" from the May 1965 edition of *Arizona Highways* magazine also features a woman in a plaid shirt and plain shorts, smiling broadly as she holds two large groups of strung-up fish (one in each hand). Unlike most other comparable images with male anglers, however, the woman does not look directly into the camera; instead, she looks over the viewer to somewhere off in the distance. Though we do not see in the image any fishing gear to suggest she was the person who caught the fish (fishing tackle, fishing hat, boat), she is nevertheless beaming with pride. Many other photographs show individuals or groups of men and women posing and grinning behind rows of bass and trout, such as in the March 1956 issue of *Arizona Highways*, where we see an image of the author, Charley Niehuis, cheerily holding up his "Lake Mead prized bass." The article goes on to describe others' tales of rainbow trout caught at the "Mecca of anglers everywhere," in which they would "laugh and shout again as [they] netted fish after fish and chalked off the wagers to be paid later . . . after the fabulous fishing was over."

FRANKLIN ROOSEVELT'S SPEECH at the dam's dedication celebration on September 30, 1935, spelled out an official purpose of the dam and forecast the significance of the project for both the region and the nation. The rationale for the dam was crystallized in Roosevelt's verbal arguments, which emphasized the need to transform natural landscapes in the interest of economic expansion and advanced his New Deal agenda and the expansion of government programs in the American West.

While much of the imagery produced subsequent to the dedication echoed the official line, some of it was quite different. In the months and years following the completion of the project, the verbal and visual discourse about the dam assumed a variety of new forms and was put to several new purposes, particularly for tourism promotion and product advertising. As the documentary photography of the construction phase gave way to a verbal and visual rhetoric of tourism and recreation, many of the sorts of images used to promote the dam as a tourist destination were utilized in advertising campaigns of various commercial products. Advertisers took advantage of the carefully crafted connotations established in earlier dam imagery in the service of selling their products.

In his dedication address, Franklin Roosevelt characterized the undeveloped dam site as a barren wasteland. In the years following the dam's completion, another strand of discourse decisively reversed that image, casting the dam and Lake Mead not only as a sublime technological wonder, but also an ideal tourist and vacation destination for boating, swimming, and fishing.

Conclusion

Descriptions of Hoover Dam have often referred to the structure as, quite literally, a monument. The July 16, 1929, *Los Angeles Examiner* pronounced the soon-to-be-built dam "a monument to the will of the people," while an article in the May 1934 edition of *Fortune* magazine commented, "The surviving monuments of a people are the labors of her builders and of her artists. . . . Out on the Colorado River, in the deep Black Canyon, America is working day and night, building herself a monument. The monument, as everyone knows, is more literally called Boulder Dam, the greatest engineering project of this day." But just as happens with other similarly iconic and symbolic man-made structures such as the Statue of Liberty or Mount Rushmore, the narrative of Hoover Dam did not end at the completion of its construction. In an article for the *Las Vegas Review-Journal* in October of 1946 — nearly a decade after the dam's completion — Florence Lee Jones reflected on what Hoover Dam had meant to those in the Southwest up to that point. She began by describing the dam as "a symbol of strength and source of power to build a gigantic war machine," and a "synonym for greatness." To her, the dam was a place where the destitute found work and a project that served as an economic salve for an impoverished citizenry. More than that, though, it was to her "a monument," a spectacle, and "proof of man's ability to harness and make gainful use of the forces of nature."

Today, the dam is still an undeniably meaningful and consequential site for many Americans. It is a secular place of public pride, what Carole Blair and Neil Michel would call a "consensual mirror of the American past and present."[1] However meaningful, though, human history is replete with cautionary tales of overinvestment in such shrines to ourselves. Goethe's classic poem *Faust* is one example that teaches us that such endeavors can lead to the worst of the seven deadly sins—pride. Goethe warns against becoming too enamored with our ability to wield technology and instrumental reasoning as weapons in the fight to overcome the limitations of nature.

In Goethe's story, the fallen angel Mephistopheles tempts Faust by placing a wager on his soul that there cannot be created an experience that will cause him to be satisfied with an earthly achievement. In spite of Mephistopheles's best attempts, Faust is unmoved by beauty, or love, or intellect. However, when Mephistopheles provides the technological means to recapture coastal lands from the sea, Faust becomes enthralled with the project and endeavors to use it to affirm the rationality of man over the disorder of the natural world. In Faust's reclamation project, then, we find that several prophetic themes emerge: the removal of recalcitrant native peoples from land they inhabit, a reliance on technology and science to achieve hegemonic power, and a divinely inspired motivation to "reclaim" the land.

Just before his death, a blind Faust revels in his earthly achievement. Filled with pride, he is delighted in the notion that his reclamation project will continue on and his utopian vision will be achieved one day when his people inhabit the reclaimed lands. Although in earlier versions of the tale, Faust (having made his pact with the devil) is then carried off to Hell, Goethe's reworking allows Faust to be saved, but only by the grace of God through the pleading of his beloved Gretchen, whose earthly life Faust had already destroyed.

The millions of dams constructed globally over the past five thousand years that humans have been manipulating waterways exemplifies our current Faustian relationship with nature. This is especially the case in the United States where reclamation projects have fundamentally altered the American landscape with over seventy-five thousand dams built in the past century alone. At the time of its construction, Hoover Dam was described as a structure of such technology and materiality that it was impervious to the oblivion of time and the ultimate victory of man over nature. The Colorado River, both revered and demonized for its impetuousness, was finally restrained to the point that, as *Popular Mechanics* described, the river was now as easy to control as turning the faucet of your own kitchen sink.

However, contrary to those who ordained the dam as destined to outlast humanity itself, we do know that it will, at some point, exceed its ability to withstand the march of time and the wear imposed by nature. Whether another dam is built in its place, new technologies allow for alternative controls of the river, or the dam is simply removed and the Colorado left to run unencumbered through Black Canyon once again, it seems ironic that the *images* of Hoover Dam and its construction—hardly thought of as having their own "life" beyond the event of their creation—will outlast those who built the dam and likely even the structure itself.

The dam's demise may, in fact, come at the behest of the American public if it decides at some future point that Hoover Dam is no longer appropriate in the zeitgeist of a more environmentally concerned society. The frenzy of dam building that the Bureau of Reclamation set into motion in the early 1900s in the United States is over, and appeals for movement in the opposite direction have been taking hold.

Bruce Babbitt's "Sledgehammer Tour" is one example of the slowly changing attitudes toward dams. On June 17, 1998, Babbitt, then secretary of the Department of the Interior, hoisted his sledgehammer and smashed it into a dam on the Menominee River, symbolically opening 160 miles of river flowing through Wisconsin and Michigan. Over the course of the next year his Sledgehammer Tour inaugurated the removal of the Quarter Neck Dam on the Neuse River in North Carolina, the Edwards Hydro Dam on Kennebec River in Maine, the McPherson Dam on Butte Creek in California, the Bear Creek Dam in Oregon, and two dams on the Elwha River in Washington, all of which allowed hundreds of miles of river to once again flow unimpeded. Now, Chinook and coho salmon, American shad, blueback herring, alewife, steelhead, striped bass, and other fish have access to thousands of miles of river habitat that had previously been blocked by dams. Removing these dams also reduced sediment, nutrient, salinity, and pollutant buildup; normalized river water temperature; returned fresh water to downstream wetlands; and improved dissolved oxygen concentrations—all problems that have been the subject of growing concern for the Colorado River, too.[2]

Recently, there has been increasing public discussion of removing another iconic dam, the Hetch Hetchy in Yosemite National Park. Hetch Hetchy provides drinking water and hydropower for over 2.5 million residents in the San Francisco Bay area, but it is one that Sierra Club founder and celebrated nature writer John Muir passionately opposed. In his 1921 book *The Yosemite,* he raged against "despoiling gain-seekers and mischief-makers" who sought to plunder sublime natural landscapes such as those in Yosemite, by "eagerly trying to make everything immediately and selfishly commercial, with schemes disguised in smug-smiling philanthropy, industriously, shampiously crying, 'Conservation, conservation, panutilization,' that man and beast may be fed and the dear Nation made great."[3]

Although writers such as Muir, Thoreau, Edward Abbey, and many others have long ago lamented the encroachment of technology upon the land and called for a more ecocentric approach to our relationship with nature, and though the political winds may slowly be changing as there exists serious discussion and actionable movement toward removing many dams, it is in fact difficult to imagine Hetch Hetchy or Hoover Dam ever being retired. California's Central Valley, which accounts for one-third of all produce grown in the United States, continues to *increase* its demands for water. Moreover, hundreds of thousands of people migrate and immigrate to the megacities of Los Angeles, San Diego, Phoenix, and San Francisco every year, with estimates of California's population alone said to increase by seven million people in the next fifteen years. This suggests that even *more* water reclamation projects will be needed, not less.

But to fully understand the desire to keep these dams in place, we need to look beyond the obvious utilitarian questions of how water distribution would continue if these mega-dams were to be removed, and instead to how and why people place value in dams not just as a functional means to direct water and provide hydropower, but as part of the way in which we understand the role of technology in our natural and cultural history. Dams and other reclamation projects represent not only natural but social control as evidenced by, for example, New Deal and TVA dam building projects as key to their attempts to wrest economic stability from the failed machinations of private capitalism. Dams have also been part of our economic and cultural landscape for hundreds of years. The first water-powered fulling mill for producing textiles was built in Virginia in 1692. Sawmills for producing lumber and gristmills for flour were abundant in the mid-1600s. These mills were significant contributors to the economic development of the early colonies and thus have become an integral part of America's heritage.

Over time, a dam and a river become a synergized cultural artifact in which dams are overlooked as technology and instead become outgrowths of the landscape. Many of those who built Hoover Dam or who captured its construction in images and prose were witnesses and chroniclers of historic change whereby the dam project fundamentally altered both the natural and cultural world. What is interesting about this process is how people have accepted these technologies as enabling the natural course of human and environmental history, and as an extension of American cultural history. However, throughout this book I have argued that we should also consider what role images have played in reifying or contesting this knowledge; the physical dam itself remains a significant rhetorical artifact, but so are its textual histories and its images.

Often the means of knowing one's cultural heritage comes by way of received familial narratives, reconstructed and representative histories, literary works, and so forth. But another manner of knowing lies in the ways that images reanimate the past by capturing a moment in time and allowing it to be passed from one place to another, from one culture to another, and from older to younger generations in a way that is extralinguistic. In other words, I do not need to be literate in the language of the culture in which an image was created in order to view the image and gain meaning from it or have an emotional response to it. Although this may seem like a profitable avenue for exploring the differences between text and image, it does, however, present a paradox of sorts in our "reading" or interpretation of an image. On the one hand, there is what has been called the rhetorical/ cultural perspective, which places significance in extracting and describing the cultural and political milieu in which an image was created, the identities and ideologies of its creators, its reception by audiences, the looking practices of the viewers, and various other attendant extra-imagistic concerns. On the other hand is what has been called the ontological perspective, which suggests that meaning lies within the content of the image itself and its cultural representations, such as in the framing of the image, the positioning and gaze of those depicted, and the linguistic signifying practices in which both we and our descriptions of the images are enmeshed.[4] But there is only so much that either of these "readings" of an image can get us. As I have attempted to demonstrate throughout this book, there is no either/ or, but a continuum that works reflexively—an infinite bridge of reciprocity that crosses from ontology to rhetoric/culture and back again.

I suggest, then, a third consideration in which images have a vitality of their own that can animate a moment for a viewer in ways that signifying practices cannot replicate and that the culture in which they were created did not intend. Images, particularly archival images such as those discussed in this book, can play an important role in both rupturing and augmenting received ways of knowing by reanimating worlds that may not be compatible with our ability—political, linguistic, or otherwise—to have described them either in the past or in the present. Images can also work to break down what Roland Barthes would call "myths," thus cleaving the ease with which we cling to the circumstance in which we find ourselves. Images do this by absorbing the context of their creation and vivifying the past for those in the future, thus ensuring that meaning is not fixed but fluid and contested. When dams, for example, become intergenerational entities understood as "natural," having been absorbed into the landscape itself and chronicled in our collective visual record for decades, images of the pre-dam landscape can bring to the fore an understanding that dams are, in fact, technologies that were constructed and that there did exist a world in which they were not a part.

IN THIS BOOK I have attempted to use the imagery of Hoover Dam to tell the story of its construction and ascension to its current place as an American cultural, visual, and technological icon. I did this by describing the variety of purposes to which the dam's imagery was put, ranging from the domination of nature to the sale of cigarettes and whiskey. My hope is that such an approach can energize a critical awareness that we should not only take into account institutional and social practices with regard to cultural artifacts, but also do so while maintaining that these practices occur against the backdrop of a physical environment. In other words, I have sought to acknowledge that nature and culture share a reciprocity of consequence; that nature complicates—in a good way—the conventional triumvirate of class, gender, and race by adding the new perspective of environment; and that hegemonic cultural structures and scopic regimes are inescapably and forcefully entangled with our culturally constructed beliefs and attitudes about nature.

But apart from these ramifications, I have also tried to show how images enliven their representations in ways that their creators might not have anticipated. Moreover, although there certainly is visual-verbal interdependence in the production of meaning, images also have an extralinguistic ability to produce degrees of knowing that words simply cannot replicate. And although understanding the cultural and historical milieu in which the images were created, and the political and ideological functions to which they were put, is an exceedingly important step that allows us to acknowledge what an image brings to us (not just what we bring to bear on the image), the process of writing this book and the experience of discovering the images have led me to believe that there is more to it than that.

Ultimately I want to understand how meaning is created through images, or what kinds of meanings might be available to certain audiences who view those images. But approaching images as a series of signs or by superimposing a linguistically based schema onto images or some other hermeneutic of convention overlooks what Keith Moxey refers to as the "exterior" of the image, "its protean interventions in the life of culture, its vitality as a representation."[5]

The images that I have described in this book interpolate the viewer; they reach out and have something to say, something that escapes time and situatedness. The images that I have been describing throughout this book are as "alive" today as they were in the moment they were taken. They have a presence that calls for us to be attuned to our own attitudes about the past, about nature, about technology, about culture, about human labor. Rather than simply relying on a rigid interpretive schema in which we pile onto images even more layers of linguistically based signifying practices, or coerce images to fit our own recognizably established patterns of meaning, my hope is that this book shows that there are opportunities to engage an openness to what the images can say *to us*. This would allow us, for example, to empathize with the lived experience of certain peoples, or relive stories that have been occluded by the state or are unable to be adequately described in words. We can look at images such as those of Hoover Dam and find that they have maintained an aura across time and culture. We can see, for example, how images that attempted to capture the essence of the technological sublime may have strayed from their originally intended political agendas, while at the same time still continuing to make meaning, all while resisting the signifying systems that we have available to describe them. Conceiving of images this way allows us to push back against our desire to capture the quintessence of an image through established systems of signification by instead looking for what Arjun Apadurai calls the "social life of things," where objects continually take on new symbolic, marketplace, and cultural values throughout the course of their existence.[6]

This is not to say that we should discount the ideological potentialities of imagery or the compelling messages that visual imagery is particularly suited to carrying. In fact, if nothing else, this book shows that in order to apprehend the meaning, or the "presence," of any particular image, it is also important to understand the circumstances of its production. The fact that some images depict scenes that the state or news media did not want people to see is evidence enough that images do have political import and do contain compelling ideological and cultural assumptions; otherwise, there would be no reason for anyone to promote authorized discourses while suppressing unauthorized ones. This also does not minimize the importance of understanding that an image frames a subject's position often at the behest of institutionally sanctioned image makers. The Bureau of Reclamation, for example, had a definite political agenda, one that is clearly and unambiguously documented in the historical record. The news media often had their own economic and political self-interests in mind when publishing images of the dam, images that were provided to them directly by the Bureau of Reclamation after careful consideration of their ability to adhere to state-sanctioned discourses. Some newspapers, such as the *Los Angeles Evening Herald,* made no attempt whatsoever to hide, and in fact were quite outspoken about their intent to be "players" and influencers in the new Southwest, of which the dam was a key part. Consequently, the process of recognizing the authorial voice of the image makers and the mediums through which those images circulated is of crucial importance if we are to apprehend how certain images make meaning.

Another concern of this book—one that is relevant in light of the traditional priorities in art history and visual studies—deals with the types of images that should be considered worthy of study. Although one of the first images in the book is from Ansel Adams, indeed a magnificent photograph, this book would have been far less compelling had it been restricted to images from Adams, Glaha, Davis, and a few other noted photographers. As this book attests, there exists outside the preserve of canonical works a vast landscape of quotidian cultural imagery that has the potential to be engaging and significant. My hope is that this book demonstrates the fruitfulness of expanding out from the "great" images that have traditionally dominated our notion of what is worthy of study, and that it provides an impetus for others to look not just to the canonical figures of a particular event or era but to other types of images that enrich our knowledge of the cultural imaginary in which they circulated.

This entire project started from an initial insight that there is no single "iconic" image of Hoover Dam, and in many ways that has been a blessing because it allowed me to freely consider brochures, illustrations, newspapers, hand-drawn renderings, book covers, and so forth. I was able to explore other images that deserve recognition without feeling the pull toward a canonized or institutionally accepted set of images deemed to have a certain aesthetic value. Moreover, this book shows that restricting oneself to only canonized imagery risks—and in fact invites—exclusion of other important images that did not make their way into wide public circulation because of ideological and institutional controls of the state or media. Indeed, these images are the very ones that provide the disruptive potential to ways of knowing that I discussed above.

Although created decades ago, and regardless of whether we consider them "art" or not, the images of Hoover Dam maintain a vivacious charisma that resists reduction to semantics, signifiers, and codes. And though their ideological potentialities have certainly changed, they freely "speak" to us with a continually evolving story to tell, one that will certainly take on new symbolic, marketplace, and cultural values over the forthcoming decades of their existence, and potentially long after the dam itself is gone.

I conclude here with a poignant observation from the May 1934 *Fortune Magazine* article in which artist Stanley Wood sought to salvage through his paintings the memory of the nameless Hoover Dam workers for future generations: "The very labor achieved effaces the laborer. . . . [Soon] the work will be finished, and the skill and speed and pride and devotion of the workers will be things of the past, and the workers themselves dispersed and forgotten. Yet not utterly forgotten. They will be remembered, if time is kind, when Boulder Dam is a mere forlorn, huge cud of stone stuck in the dry stone gut of a forgotten river. Remembered in the gray-white record of a lens, and in the skillful assemblage of colors on paper."

Notes

Introduction

1. Hiram W. Johnson, "The Boulder Canyon Project."

2. The first was a three-cent stamp issued in September 1935—the same month as the official dedication of the dam—that went on to sell over seventy-three million copies, while the second currently has the designation of being the highest denomination ever issued by the Postal Service at $16.50, an express-mail, limited-edition stamp from June of 2008.

3. Louis Lozowick, *William Gropper*.

4. These range from civil, structural, and electrical engineering; to geography, water resources, and hydrography; to the history of technology, environmental history, and oral history, not to mention the hundreds of books, book chapters, and newspaper and journal articles that discuss dams and water resources as integral to the history and future of the American West.

5. Barbara Ann Vilander, in her dissertation "The Hoover Dam Photographs of Ben Glaha Completed Under the Auspices of the United States Bureau of Reclamation," presents a detailed account of the work of celebrated Bureau of Reclamation photographer Ben Glaha. She later adapted her dissertation into the book *Hoover Dam: Photographs of Ben Glaha*. Richard Guy Wilson has written extensively about the architectural design of the dam, work that has involved some consideration of visual representations. See Richard Guy Wilson, "Massive Deco Monument," 45–47; Richard Guy Wilson, "Machine-Age Iconography in the American West: The Design of Hoover Dam," 463–93; Richard Guy Wilson, "American Modernism in the West: Hoover Dam," 291–319. I have located no other studies that treat visual images of Hoover Dam at any length.

6. To be clear, I use the terms *icon* and *iconic* throughout this book not in any specific sense, but in reference to how the dam is treated as an American cultural icon. My aim is to explore how the dam came to be what might be called an image icon, with that term driven mostly by its vernacular usage in which an image (or in this case a collection of images) takes on culturally and historically constructed symbolic meaning outside of its individual components. Image icons such as Hoover Dam, the Statue of Liberty, and Marilyn Monroe, once removed from their immediate context, take on symbolic adherents that are perceived to be universal (industry, freedom, sexuality) but are in fact representative of particular political, ideological, and cultural traditions. The dam, then, is iconic in the sense that its images have been appropriated countless times, it has symbolic meaning for many people, and it is both an icon of industry and an icon of commodity, one that has been mass-produced and marketed.

7. While I do treat both mechanically produced images (photographs) and images composed by human hand, I acknowledge that there are significant differences

between these two orders of visual representation, most notably in relation to the temporality of depiction. Whereas a photograph's chief effect is as record of the visible past, other forms of representation not so tied to any particular moment in real time (such as illustrations) can more readily suggest the future, and in some cases a wildly imaginative future using projective conceptions of an imagined object-to-be.

8. Robert Hariman and John Louis Lucaites, in their important and influential work *No Caption Needed: Iconic Photographs, Public Culture, and Liberal Democracy*, center their attention on how iconic photographs work to construct civic identity. They define iconic photographs as "images appearing in print, electronic, or digital media that are widely recognized and remembered, are understood to be representations of historically significant events, activate strong emotional identification or response, and are reproduced across a range of media, genres, or topics." See also Robert Hariman and John Louis Lucaites, "Performing Civic Identity: The Iconic Photograph of the Flag Raising at Iwo Jima."

9. Roland Barthes, *Camera Lucida: Reflections on Photography*, 6.

10. It is vitally important, however, to make a distinction at the outset between what can and cannot be accomplished in any analysis that considers the environ- ment. As discussed by Lawrence Buell and others, a central concern of any consideration of nature or environment is the fundamental problem of humans being able to speak *on behalf of* nature. People cannot speak from the experience of *being* nature like a woman would for feminism, for example. One can speak in the interests of nature, but no one can claim that they are speaking *as* nature. Consequently, what differentiates such ecocentric perspectives from other discourses speaking at the behest of silenced or disempowered groups is that the "Other" in this case is nature and is something outside of both culture and the human body. See Lawrence Buell, "Letter"; and Lawrence Buell, *The Future of Environmental Criticism: Environmental Crisis and Literary Imagination*.

11. High modernism primarily developed as a social theory to describe how governments use regulations and scientific theories to order the social world by "sedentarizing" the masses and by streamlining traditional state administrative functions in the search for greater efficiency, control, and power. In fact, to Scott, the most egregious developments in modern state-sponsored social engineering occurred as a result of unrestrained uses (or abuses) of power that pushed a rigorously controlled "ordering of nature and society," while attending to a "prostrate civil society that lacks the capacity to resist these plans." See James C. Scott, *Seeing Like a State: How Certain Schemes to Improve the Human Condition Have Failed*, 88–89.

12. The terms *visual studies*, *visual culture*, and *visual cultural studies* are often used interchangeably to describe the endeavor of exploring the visual and visuality as constructed in media, the arts, and increasingly in the quotidian. However, W. J. T. Mitchell in "Showing Seeing: a Critique of Visual Culture" posits a differentiation between those terms. He writes,

I think it's useful at the outset to distinguish between visual studies and visual culture as, respectively, the field of study and the object or target of study. Visual studies is the study of visual culture. This avoids the ambiguity that plagues subjects like history, in which the field and the things covered by the field bear the same name. In practice, of course, we often confuse the two, and I prefer to let visual culture stand for both the field and its content, and to let the context clarify the meaning. I also prefer visual culture because it is less neutral than visual studies, and commits one at the outset to a set of hypotheses that need to be tested—for example, that vision is (as we say) a cultural construction, that it is learned and cultivated, not simply given by nature; that therefore it might have a history related in some yet to be determined way to the history of arts, technologies, media, and social practices of display and spectatorship; and (finally) that it is deeply involved with human societies, with the ethics and politics, aesthetics and epistemology of seeing and being seen.

13. Margaret Dikovitskaya, *Visual Culture: The Study of the Visual After the Cultural Turn*. Nicolas Mirzoeff also describes visual culture as highlighting "those moments where the visual is contested, debated and transformed as a constantly challenging place of social interaction and definition in terms of class, gender, sexual and radicalized identities." He also describes it as a "resolutely interdisciplinary subject," the focus of which "crosses the borders of traditional academic disciplines at will." As a result, he claims unapologetically that visual culture, as an area of study, is not an academic discipline but an approach or tactic. "It is a fluid interpretive structure," he writes, "centered on understanding the response to visual media of both individuals and groups." See Nicolas Mirzoeff, *An Introduction to Visual Culture*, 4. For an overview of theoretical perspectives and debates within the expanding field of visual studies/ culture, see also Norman Bryson, Michael Ann Holly, and Keith Moxey, eds., *Visual Culture: Images and Interpretations*.

14. Martin Jay, for example, has established the term *ocularcentrism* to describe the centrality of vision and the visual in modern Western societies and argues that, after a considerable period in which the visual was subordinated to verbal text, the eighteenth century witnessed a growing dependence on visual images, particularly with regard to scientific investigation. See Martin Jay, *Downcast Eyes: The Denigration of Vision in Twentieth-Century French Thought*.

15. Roland Barthes, for example, writes in "The Rhetoric of the Image" that although we live in a culture saturated with visual imagery, these images are almost always accompanied by words. To him, "It is not very accurate to talk of a civilization of the image—we are still, and more than ever, a civilization of writing, writing and speech continuing to be the full terms of the informational structure." Stuart Hall, too, has argued that although photographs add new dimensions to what is communicated in modern newspapers, for example, text is still the essential element while the photograph remains the optional one. Despite the fact that these particular commentaries appeared prior to the digital revolution, the remarks of Barthes and Hall are still relevant and should continue to command attention. See Roland Barthes, "The Rhetoric of the Image"; Stuart Hall, "The Determinations of News Photographs."

16. See Cara A. Finnegan and Jennifer L. Jones-Barbour, "Review Essay: Visualizing

Public Address," in which they address issues of rhetorical context and rhetorical circulation in several recent works of visual rhetorical scholarship.

17. See, for example, Jefferson Hunter, *Image and Word: The Interaction of Twentieth-Century Photographs and Texts.*

18. For book-length discussions of images and ideology, see W. J. T. Mitchell, *Picture Theory: Essays on Verbal and Visual Representation;* and W. J. T. Mitchell, *Iconology: Image, Text, Ideology.*

19. Donna J. Haraway, *Simians, Cyborgs, and Women: The Reinvention of Nature.*

20. Jean Baudrillard, *America,* 77.

1 / Nature, Culture, and the Divine Right of Transformation

1. Robert Young, *Colonial Desire,* 23–25.

2. However, Tylor's faith in "race sciences"—which posits that no significant changes in humans have occurred in thousands of years, thus different races and societies have been concretized within their pecking order of cultural and intellectual attainment—is clearly problematic. To escape the racial baggage of these anthropological foundations, people like Stuart Hall and Terry Eagleton argue for a political undercurrent to cultural practice in which people create, define, and express their identity within, and as part of, a community. Although this notion of culture brings with it some of the patrimonial associations with race and racism that we cannot evade simply by suggesting that we are now somehow post-racial, a focus on the *how* and *why* instead of *better* and *worse* allows distinctions between groups while attempting to avoid racialized cultural interpretations. See Mirzoeff, *Introduction to Visual Culture.*

3. Fredrick Jackson Turner's 1893 "The Significance of the Frontier in American History" is often credited with shaping early notions of the American West and Americans' pioneering characteristics of rugged individualism, but recent scholarship, like that of Patricia Nelson Limerick, Richard White, William Cronon, and Donald Worster, highlights the exploitation of Western natural resources (sometimes referred to as "New Western History") while challenging the traditional centrality of Turner's Western ideal.

4. For the influence of Fredrick Jackson Turner's thesis on historians and subsequent notions regarding the American West, see William Cronon, "Revisiting the Vanishing Frontier: The Legacy of Frederick Jackson Turner."

5. Theodore Steinberg, "Fertilizing the Tree of Knowledge: Environmental History Comes of Age."

6. Hal Rothman, "Conceptualizing the Real: Environmental History and American Studies."

7. Donald Worster's excellent work has ranged from exploring the intellectual connection between science and nature, in which he engages environmental history starting from a "pure" time in which humanity embraced agriculture and was able to limit its impact on the land, to modern ecological disasters and the hierarchical societies they produced that were the consequence of cultural social choices and human

actions as much as they were from environmental circumstances. Donald Worster, *Rivers of Empire: Water, Aridity, and the Growth of the American West*; Donald Worster, *Nature's Economy: A History of Ecological Ideas.*

8. William Cronon, "The Trouble with Wilderness; or, Getting Back to the Wrong Nature."

9. Neil Evernden, *The Social Creation of Nature.*

10. William Cronon is probably the foremost contemporary authority in arguing this point.

11. William Cronon, *Nature's Metropolis: Chicago and the Great West*, xvi.

12. Ibid., 17. See also William Cronon, *Changes in the Land: Indians, Colonists, and the Ecology of New England.*

13. Terry Eagleton, *The Idea of Culture*, 2.

14. Ibid., 67.

15. Ibid., 110.

16. Ibid., 122.

17. Ibid., 68.

18. Ibid., 69.

19. Kevin Starr, *Material Dreams: Southern California Through the 1920s*, 20; Worster, *Rivers of Empire*, 200.

20. "Hoover Dam No Longer Dream as Drills Hum," *Las Vegas Evening Review-Journal*, January 15, 1932.

21. The Great Depression would lead to conditions that would motivate scores of unemployed and downtrodden people to migrate to Hoover Dam in search of work once construction started. See Theodore Steinberg, "'That World's Fair Feeling': Control of Water in 20th-Century America."

22. Carolyn Merchant, *Reinventing Eden: The Fate of Nature in Western Culture.*

23. Ibid., 20.

24. It should be noted that there has been much debate in environmental history, environmental studies, and ecocriticism regarding the term *nature*. See, for example, Timothy Morton, *Ecology Without Nature: Rethinking Environmental Aesthetics*; Daniel J. Philippon, *Conserving Words: How American Nature Writers Shaped the Environmental Movement*, 9–16.

25. Philippon, *Conserving Words*, 6.

26. Starr, *Material Dreams*, 20.

27. R. G. Skerrett, "America's Wonder River—the Colorado," 12.

28. Ibid., 13.

29. Leo Pasvolsky, "Many Problems to Be Solved in Harnessing the Colorado River," *New York Tribune*, September 10, 1922.

30. Robert L. Duffus, "Vast Energy Locked in River: Six Million Horsepower Waiting for Release in Colorado," *New York Times*, May 11, 1924.

31. For commentary on the Garden as a second creation theme, see Thomas P. Hughes, *Human-Built World: How to Think About Technology and Culture*; David E. Nye, *America as Second Creation: Technology and Narratives of New Beginnings*; Merchant, *Reinventing Eden.*

32. James G. Cantrill and Christine L. Oravec, eds., *The Symbolic Earth: Discourse and Our Creation of the Environment.*

33. This notion continued into the nineteenth century. Daniel Rogers writes of nineteenth-century American laborers that there was "a sociology to the work ethic as well as an amalgam of ideas. Praise of work . . . was strongest among the middling, largely Protestant, property-owning classes: farmers, merchants, ministers, and professional men, independent craftsmen, and nascent industrialists. Such groups had formed the backbone of English Puritanism." See Daniel T. Rodgers, *The Work Ethic in Industrial America, 1850-1920,* 14.

34. Lee Braude, *Work and Workers: A Sociological Analysis.*

35. Seymour Martin Lipset, "The Work Ethic, Then and Now."

36. Nye, *America as Second Creation,* 9-10; Henry Nash Smith, *Virgin Land: The American West as Symbol and Myth,* 125-26.

37. Thomas Jefferson, "To John Jay, 1785."

38. Thomas Jefferson, "To George Washington, 1787."

39. In the section titled "Rights, Private and Public: Lands" Jefferson writes, "Every person of full age neither owning nor having owned 50 acres of land, shall be entitled to an appropriation of 50 acres or to so much as shall make up what he owns or has owned 50 acres in full and absolute dominion. And no other person shall be capable of taking an appropriation." For commentary on Jefferson's agrarianism, see also Howard Gillman, *The Constitution Besieged: The Rise and Demise of Lochner Era Police Powers Jurisprudence,* 25-26.

40. Thomas Jefferson, *The Papers of Thomas Jefferson,* vol. 1, 1760-1776, 539.

41. Thomas Jefferson, "Notes on the State of Virginia: Query XIX."

42. Smith, *Virgin Land,* 129.

43. Leo Marx, *The Machine in the Garden: Technology and the Pastoral Ideal in America.*

44. Ibid., 88.

45. Ibid.

46. For a discussion of pastoralism and its significance for environmentalism, see Greg Garrard, *Ecocriticism,* 33-58.

47. Capitalizing on high-profile projects such as Hoover Dam, and riding a wave of governmental and popular approval, the Bureau of Reclamation issued in 1946 its encyclopedic *The Colorado River: A Comprehensive Report on the Development of the Water Resources of the Colorado River Basin for Irrigation, Power Production, and Other Beneficial Uses in Arizona, California, Colorado, Nevada, New Mexico, Utah, and Wyoming,* outlining 134 water projects intended to develop the entire upper Colorado River basin.

48. Kendrick A. Clements, "Herbert Hoover and Conservation 1921-33."

49. Quoted in Whitney R. Cross, "Ideas in Politics: The Conservation Policies of the Two Roosevelts," 425.

50. Gertrude Almy Slichter, "Franklin D. Roosevelt and the Farm Problem, 1929-1932," 242.

51. Robert L. Duffus, "Four Vast Power Projects to Make Economic History," *New York Times,* October 2, 1932.

52. The notion of dominating (the ascendancy of the power of man, machine, and technology to seize control over nature) the Colorado River in the service of expanding the economy has roots that go back through, and far beyond, John Wesley Powell and the first European pioneers to travel into the American West.

53. Marx, *Machine in the Garden*, 194.

2 / Natural Disasters and Political Adversity

1. Philip L. Fradkin, *A River No More: The Colorado River and the West*, 15.

2. For an introduction to the Great Depression and its causes and outcomes, see Michael A. Bernstein, *The Great Depression: Delayed Recovery and Economic Change in America, 1929–1939*; Thomas E. Hall and J. David Ferguson, *The Great Depression: An International Disaster of Perverse Economic Policies*; David M. Kennedy, *Freedom from Fear: The American People in Depression and War, 1929–1945*; Eric Rauchway, *The Great Depression and the New Deal: A Very Short Introduction*.

3. The Imperial Valley was also the subject of popular books and movies in the 1920s. *The Winning of Barbara Worth* (1911), a wildly popular best-selling novel written by Harold Bell Wright (later made into a film in 1926 starring Ronald Colman and a then-unknown Gary Cooper), tells the story of a reclamation project in the Imperial Valley that closely follows actual events that led to consideration of Hoover Dam. In the story, rancher Jefferson Worth finds Barbara, a four-year-old whose family is killed in a sandstorm, in the desert and raises her as his own daughter. Later, Barbara envisions bringing water to the valley to create fertile fields supporting hundreds of families who farm the land. However, her plans are spoiled when the antagonist of the story, James Greenfield (a New Yorker), begins his own profiteering reclamation project by building a dam without regard to the safety or welfare of the families in the valley. When Greenfield's dam and levees burst, issuing a torrential flood that threatens the lives of everyone in the valley, the hero Willard Holmes, who embodies the Western code of honor, secures the river by replacing the substandard material used by Greenfield with quality material provided by the Worth Company. After Greenfield's corporation abandons the valley, the Worth Company takes control of the project and oversees the development of the farmland. This novel (a 511-page epic) sold over two million copies by 1920 and has sold over nine million copies to date. For details on the extraordinary success of this novel in the early 1900s, see Margaret Macmullen, "Love's Old Sweetish Song." Gerry Chudleigh has compiled an extensive Website dedicated to the writings of Harold Bell Wright, including a section on *The Winning of Barbara Worth*; see http://www.hbw.addr.com.

4. Worster, *Rivers of Empire*, 34. For discussions of indigenous people's water projects, see also Fradkin, *River No More*, 17–23.

5. Donald C. Jackson, "Origins of Boulder/Hoover Dam: Siting, Design, and Hydroelectric Power."

6. Joseph E. Stevens, *Hoover Dam: An American Adventure*, 7.

7. Norris Hundley Jr., *The Great Thirst: Californians and Water—a History*, 205.

8. Frederick D. Kershner Jr., "George Chaffey and the Irrigation Frontier."

9. Starr, *Material Dreams*, 25–29.

10. Stevens, *Hoover Dam*, 14.

11. Presidents Roosevelt and Taft both recommended to Congress that Harriman be reimbursed for his efforts, but Congress never acted on these requests. For a detailed account of the flood of 1905, see Starr, *Material Dreams*, 21–41.

12. Ibid., 10. See also Worster, *Rivers of Empire*, 197. For an account of the creation and subsequent use of the Salton Sea, see Pat Laflin, *The Salton Sea: California's Overlooked Treasure*.

13. Beverley B. Moeller, *Phil Swing and Boulder Dam*, 10–11.

14. Worster, *Rivers of Empire*, 201.

15. Ibid., 206.

16. Sherman Buck, "Letter to Jennie Surbaugh Buck, March 8, 1924," Richard Mercurio Family Archive, Valley Center, CA.

17. Donald J. Pisani, *Water and American Government: The Reclamation Bureau, National Water Policy, and the West, 1902–1935*.

18. Worster, *Rivers of Empire*, 202. See also Steinberg, "'That World's Fair Feeling.'"

19. Many books have been written on this topic. For a few treatments of the political struggles over water rights in the Southwest, see William L. Kahrl, *Water and Power: The Conflict over Los Angeles Water Supply in the Owens Valley*; Donald J. Pisani, *Water, Land, and Law in the West: The Limits of Public Policy, 1850–1920*.

20. See also Paul R. Josephson, *Industrialized Nature: Brute Force and the Transformation of the Natural World*; Patricia Nelson Limerick, *The Legacy of Conquest: The Unbroken Past of the American West*; William H. Goetzmann, *Exploration and Empire: The Explorer and the Scientist in the Winning of the American West*.

21. Richard Guy Wilson writes, "The individual responsible for the conception of the project in 1902 and directing the early investigations was Arthur Powell Davis, a nephew of the first explorer of the Colorado River, John Wesley Powell. Davis was an engineer and a director of the Reclamation Service in the 1910s and 1920s. Although he had left the agency in 1923, his ideas formed the basis of the design. Elwood Mead, the commissioner of reclamation during most of the construction period and for whom Lake Mead—the reservoir—was named, approved the plans." R. G. Wilson, "Machine-Age Iconography in the American West," 474–76.

22. John Whipple and Estevan López, "New Mexico's Experience with Interstate Water Agreements."

23. The Arizona legislature initially refused to ratify the compact and held out for over two decades. It finally relented, however, in signing on to the seven-state compact on February 24, 1944.

24. "Renew Fight Today on Boulder Dam Bill," *New York Times*, February 24, 1927.

25. Ibid., 4.

26. Ibid.

27. Pasvolsky, "Many Problems to Be Solved."

28. Johnson, "Boulder Canyon Project," 150.

29. Duffus, "Vast Energy Locked in River."

30. "Arizona First and Last," *Coconino Sun*, November 25, 1921.

31. James B. Girand, "Pioneer in Planning Colorado River Development, Declares Arizona Must Guard Against California," *Bisbee Daily Review*, June 28–29, 1922.

32. "McGregor Warns of Dangers of Swing-Johnson Bill to Arizona," *Coconino Sun*, October 27, 1922.

33. See "Arizona and the Colorado," *Arizona Daily Star*, January 16, 1944; Marc Reisner, *Cadillac Desert: The American West and Its Disappearing Water*, 255–59.

34. "Can They Beat the Federal Government on Boulder Dam?," *Mohave County Miner and Our Mineral Wealth*, July 14, 1922, 8.

35. "Charges Huge Lobby Against Hoover Dam," *New York Times*, November 11, 1927.

36. Ernest Gruening, "Power and Propaganda."

37. Stevens, *Hoover Dam*, 18, 26–27, 32.

38. See "So Mr. Ickes Was Wrong," *San Francisco Chronicle*, January 29, 1939; "'Honest' Harold Ickes Stewing Again," *Newark Star-Eagle*, March 1, 1939; Edward Ainsworth, "Why Mr. Ickes!," *Los Angeles Times*, January 30, 1939; Boulder City Museum and Historical Association Website: www.bcmha.org.

39. The Six Companies organization also went on to build Parker Dam and the Grand Coulee Dam, laid the foundations for the Golden Gate Bridge, and would build ships and contract for various construction projects for the US military during World War II. It would go on to dominate the steel, aluminum, concrete, chemical, and automotive industries and later become the largest engineering firm in the world. Names such as Warren A. Bechtel, Henry J. Kaiser, Charlie Shea, and Felix Kahn were all part of the original Six Companies Inc.

40. For a review of legal cases involving Arizona and Colorado River water and Hoover Dam from the 1920s through the 1960s, see Frank J. Trelease, "Arizona V. California: Allocation of Water Resources to People, States, and Nation."

3 / Illustrating the Dam

1. Robert D. McCracken, *Las Vegas: The Great American Playground*, 33.

2. Paul Messaris, *Visual Literacy: Image, Mind, and Reality*, 46.

3. E. H. Gombrich, *Art and Illusion: A Study in the Psychology of Pictorial Representation*, 33–62. For additional commentary on Gombrich and what W. J. T. Mitchell calls the distinction between "natural" and "conventional" signs, see W. J. T. Mitchell, *Iconology*, 75–94. See also David Marr, *Vision: A Computational Investigation into the Human Representation and Processing of Visual Information*, in which he theorizes the transformations that visual information undergoes in order for the brain to construct a sense of the three-dimensional world.

4. Messaris, *Visual Literacy*, 49.

5. Ibid., 50. I am appropriating the term *pictorially inexperienced* and implying it in a slightly different way than Messaris; however, the difference bears no consequence to my argument. I use the term to suggest that viewers of Hoover Dam imagery had not yet seen the real-world object (as it was not built yet), only representations in newspapers, pamphlets, or other media, some of which would have been first-time

encounters with representational imagery of the dam. Messaris's notion of a "pictorially inexperienced" person is described within the context of a larger study of anthropologic research into comprehension of images by persons who have *never seen a reproduced image* (such as a drawing or photograph). Messaris describes one study in which individuals were given photographs of members of their tribe. The viewers did not make the connection between the representation of the person in the image and the actual tribal member. However, Messaris points out that, rather than a commentary on their inability to connect representative images to real-world objects, this outcome could have been a result of the viewer never encountering paper before and being, instead, preoccupied with comprehending the medium.

6. Ibid., 51.

7. Ibid., 53.

8. Ibid., 54–55.

9. Ibid., 55.

10. Ibid., 58.

11. I think it is fair to assume that the Washington Monument is an object with which readers might have had basic familiarity, even if a significant number of the citizens of Lytton, Iowa, had not visited the nation's capital to view the monument in person.

12. Arthur P. Davis is also the "Davis" in the Fall-Davis report entitled "Problems of Imperial Valley and Vicinity," which was prepared under the Kinkaid Act and submitted to Congress on February 28, 1922. The report recommends constructing the All-American Canal and a "high dam on the Colorado River at or near Boulder Canyon." Later that year, on November 24, 1922, representatives of the Colorado River basin states signed the Colorado River Compact in Santa Fe, New Mexico (sometimes referred to as the Santa Fe Pact). Shortly thereafter, the first of the Swing-Johnson bills to authorize a dam and canal was introduced in Congress.

13. For his additional writings on the Boulder Dam project, see E. F. Scattergood, "Engineering and Economic Features of the Boulder Dam."

14. John Berger, *Ways of Seeing*, 7.

15. Aside from the short account of the architecture of the dam written by the chief architect himself in Gordon B. Kaufmann, "The Architecture of Boulder Dam," the most insightful and comprehensive analysis of Hoover Dam's design is R. G. Wilson, "Machine-Age Iconography in the American West." Wilson's study details the history of the decisions and the personalities that led from the initial design ideas, through modifications made in the 1920s and early 1930s, to the completed technical design specifications. See also R. G. Wilson, "American Modernism in the West."

16. Kaufmann received the commission for the Los Angeles Times Building in 1931 from newspaper publisher Harry Chandler, who was also a strong supporter of the dam project. It is not known whether this connection had any influence on Kaufmann being selected to consult on the Hoover Dam project.

17. R. G. Wilson, "Machine-Age Iconography in the American West," 482.

18. R. G. Wilson, "American Modernism in the West," 302–3.

19. Ibid., 304–5.

20. Kaufmann, "Architecture of Boulder Dam."

21. Ibid., 3.

22. R. G. Wilson, "Machine-Age Iconography in the American West," 466–67.

23. Baudrillard, *America*, 90.

24. Kevin Hetherington, *Badlands of Modernity: Heterotopia and Social Ordering*.

25. Carole Blair, Marsha S. Jeppeson, and Enrico Pucci Jr., "Public Memorializing in Postmodernity: The Vietnam Veterans Memorial as Prototype."

26. Section 4(a) of the Boulder Canyon Project Act of 1928 was authorized on only one of the two contingencies: (1) if all seven states ratified the Colorado River Compact, or (2) if six would ratify and California would pass a law limiting its water use to 4.4 million acre-feet per year out of the 7.5 million acre-feet total apportionment to the lower basin states, plus one half of any outstanding unapportioned excess or surplus water. See Trelease, "Arizona V. California," 173.

27. "Loops" or "loopholes" in medieval castles were narrow openings in the walls through which light and air could come in and archers could shoot out.

28. Marita Sturken and Lisa Cartwright, *Practices of Looking: An Introduction to Visual Culture*, 22–24.

29. Hon. John W. Summers, "Jack Rabbits and Markets," *Congressional Record*, Speech to Sixty-Seventh Congress, Second Session, June 3, 1922.

30. Ibid.

31. J. B. Harley, "Deconstructing the Map," 4.

32. Alan M. MacEachren, *How Maps Work: Representation, Visualization, and Design*, 12.

33. With regard to maps and mapmaking generally, this is a position that has, until very recently, been ignored by professional cartographers and geographers. For detailed discussions of cartography's historically positivist epistemology, see MacEachren, *How Maps Work*; and Harley, "Deconstructing the Map."

34. Ben F. Barton and Marthalee S. Barton, "Ideology and the Map: Toward a Postmodern Visual Design Practice," 233.

35. For an analysis of maps as a form of visual rhetoric, see Amy Propen, "Visual Communication and the Map: How Maps as Visual Objects Convey Meaning in Specific Contexts."

36. Barton and Barton, "Ideology and the Map," 239.

37. Harley, "Deconstructing the Map," 9–10.

38. Barton and Barton, "Ideology and the Map," 236.

39. Michel de Certeau, *The Practice of Everyday Life*, 113.

40. Transportation imagery in maps of the dam is a prominent and ongoing motif in both pre-construction- and construction-phase imagery. In the *Lytton Star* of October 6, 1921, a map of the Southwest indicating the future site of the dam sits below an image of three men on horseback. The *Examiner* map includes cargo ships, cruise liners, futuristic electric trains, tractors, and airplanes. And, as we will see in subsequent chapters, maps are invariably part of the pictorial montage of newspaper

travelogue articles about automobile trips to the dam site. Transportation is also frequently visually linked in documentation of the dam's construction, advertisements for cars or trucks, tourism promotion, views of the dam showing cars driving across the dam, and so forth.

41. de Certeau, *Practice of Everyday Life,* 121.

42. Ibid., 113.

43. Marx, *Machine in the Garden,* 141. For additional commentary on the railroad in American art, see Susan Danly and Leo Marx, eds., *Railroad in American Art: Representations of Technological Change.*

44. Barton and Barton, "Ideology and the Map," 236.

45. Ibid., 236.

4 / Picturing the Sublime

1. Stevens, *Hoover Dam,* 265–66.

2. Daniel Rosenberg, "No One Is Buried in Hoover Dam: Modernism and Rumor," 87.

3. See, for example, University of California, Los Angeles, Film and Television Archive, Hearst Metronome News, "Boulder Dam" (1933–49), VA4884; "Work Starts on Hoover Dam" (vol. 1, no. 303, September 24, 1930), VA13886; "Giant Boulder Dam Springs to Life" (vol. 7, no. 303, September 14, 1936), VA12710. International Newsreel, "Airplane Survey Shows Impressive Magnitude of the Boulder Dam Project" (vol. 11, no. 9, January 29, 1929), VA12741. For more on the filmed documentation of the construction project, see R. G. Wilson, "American Modernism in the West," 467, 474, 493, with notes.

4. See Hearst Metronome News, "Mormon Choir of 450 Voices of Salt Lake City Tabernacle Singing atop Boulder Dam" (vol. 6, no. 289, July 29, 1935), VA12749.

5. Copeland Lake, "Huge Trailer Hauls Penstock Pipe at Boulder Dam," 129–30.

6. Instead of describing the "deep structures" of an open and total language system, its reliance on signs instead necessitated contingency and historical context for any understanding of culture, which was, in practice, a closed system in which meaning's ability to cross time and place was impossible.

7. Gunther Kress and Theo van Leeuwen, *Reading Images: The Grammar of Visual Design,* for example, is an attempt to lay the groundwork for just such an education in visual literacy.

8. Mirzoeff, *Introduction to Visual Culture,* 14.

9. R. G. Wilson, "Machine-Age Iconography in the American West," 493.

10. Reisner, *Cadillac Desert,* 129.

11. Gregory Clark, *Rhetorical Landscapes in America: Variations on a Theme from Kenneth Burke.*

12. David E. Nye, *American Technological Sublime,* 16.

13. For a discussion of mass representations of the sublime through Ansel Adams's photographs, see Nathan Stormer, "Addressing the Sublime: Space, Mass Representation, and the Unpresentable."

14. Christine L. Oravec, "The Sublime."

15. As Philip Shaw in *The Sublime* points out, Longinus's account differed from previous rhetorical treatises because sublimity, although an effect of fine oratory, did not have the same pedagogical status as other standard rhetorical canons of *inventio, dispositio, elocutio, memoria*, and *actio*.

16. Longinus, *On Great Writing (On the Sublime)*, I.4.

17. The masculine gendered language here is deliberate, as I discuss throughout this book the heroic and decidedly male figure of the engineer and the hypermasculinized imagery and language that are attached to the Hoover Dam project.

18. Immanuel Kant, *Observations on the Feeling of the Beautiful and Sublime*, 91.

19. Ibid., 92.

20. For a summary of these positions, see Stormer, "Addressing the Sublime," 215–16.

21. Ibid., 216.

22. Jean-François Lyotard, "The Sublime and the Avant-Garde," 93.

23. Nye, *American Technological Sublime*, xiii–xiv. Nye was not the first person to use the term *technological sublime*. Perry Miller is thought to have coined the term in his book *The Life of the Mind in America: From the Revolution to the Civil War*. John F. Kasson then took up and expanded the concept in his seminal work *Civilizing the Machine: Technology and Republican Values in America, 1776–1900*. Although many others have considered aspects of the technological sublime, I draw primarily on Nye because instead of examining notions of aesthetic experience as conceived in literary theory, poetics, American landscape painting, or the shifting definitions of the sublime over the past two thousand years, his work centers on "experiences which ordinary people have intensely valued . . . repeated experiences of awe and wonder, often tinged with an element of terror, which people have had when confronted with particular natural sites, architectural forms, and technological advancements" (xvi–xvii). For other considerations of the technological sublime, see Roland Marchand, *Advertising the American Dream: Making Way for Modernity, 1920–1940*; Barbara Novak, *American Painting of the Nineteenth Century: Realism, Idealism, and the American Experience*; John F. Sears, *Sacred Places: American Tourist Attractions in the Nineteenth Century*.

24. Nye, *American Technological Sublime*, xvi–xviii.

25. Theophilus Ballou White, "Building the Big Dam."

26. Susan Stewart, *On Longing: Narratives of the Miniature, the Gigantic, the Souvenir, the Collection*, 44.

27. Jacques Derrida, *The Truth in Painting*.

28. Stormer, "Addressing the Sublime," 217.

29. R. Robert Russell, "Highlights on Boulder Dam."

30. Florence Lee Jones, "Spectacle of Beauty; Pride of World Unknown Thousands Helped Build," *Las Vegas Evening Review-Journal*, October 22, 1946.

31. Muscle Shoals is a city on the Tennessee River in northwest Alabama and the location of Wilson Dam. President Woodrow Wilson authorized the dam as part of a plan to produce nitrates for explosives for use in World War I, although construction

did not begin until two years after the war, in February of 1918. Henry Ford offered to buy the dam for $5 million ($41.5 million less than the cost of construction) as part of a plan to build a factory in Muscle Shoals, but Congress rejected his offer and Muscle Shoals/Wilson Dam became the catalyst for the formation of the Tennessee Valley Authority. Dnieperstroy (or the Dnieper Dam) is a hydroelectric dam on the Dnieper River in Zaporizhia, Ukraine (formerly Aleksandrovsk, Russia). At the time of its completion in 1932 it was one of the largest dams in the world and was promoted by Stalin as a triumph of the Bolsheviks. The dam was dynamited twice in World War II, once by retreating Russian troops in 1941 and again by retreating German troops in 1943. It was rebuilt after the war and still stands today.

32. Russell, "Highlights on Boulder Dam," 4.

33. Dorothy Childs Hogner, "Boulder Dam," 5–7, 33, 34.

34. Oskar J. W. Hansen, "With the Look of Eagles." Hansen was selected by project architect Gordon B. Kauffman and was appointed consulting sculptor of the Bureau of Reclamation by Secretary of the Interior Harold Ickes, after a national competition in search of suitable designs for the dam. Hansen is responsible for the winged sculptures, the ten bas-reliefs on the elevator shafts, the flagpole and star chart, and several memorial plaques at the dam. For detailed discussion of the architecture and artwork at the dam, see Kaufman, "Architecture of Boulder Dam"; R. G. Wilson, "Machine-Age Iconography in the American West"; R. G. Wilson, "Massive Deco Monument." For a discussion of how architecture works rhetorically as language, movement, and fantasy, see Darryl Hattenhauer, "The Rhetoric of Architecture: A Semiotic Approach."

35. This sentiment is repeated word for word in the article "Boulder Dam—Lake Mead" in the August 1945 edition of *Nevada Magazine*.

36. Hansen, "With the Look of Eagles," 2.

37. According to Hansen, "Instead of measuring 12 earthly hours this dial measures a *Platonic Year*. A *Platonic Year*, or *Great Year*, according to Stockwell, is made up of 25,694.8 of our ordinary years. We cannot be sure of the exact length of the *Platonic Year*, because the civilized history of man and hence astronomy is at the most only 10,000 years old; but we feel certain that it may not vary from the above mean time by more that 281.2 of our ordinary years."

38. Oskar J. W. Hansen, "A Split Second Petrified on the Face of the Universal Clock."

39. Ibid., 5.

40. Ibid., 6.

41. The best analysis of Niagara Falls as both American and natural sublime is Elizabeth R. McKinsey, *Niagara Falls: Icon of the American Sublime*.

42. For discussions of the American sublime in literature, see Harold Bloom, "Emerson and Whitman: The American Sublime"; David S. Miall, "Foregrounding and the Sublime: Shelley in Chamonix"; Rob Wilson, *American Sublime: Genealogy of a Poetic Genre*. For discussion of the American sublime in art, see Nicholas Green, *The Spectacle of Nature: Landscape and Bourgeois Culture in Nineteenth-Century France;*

Andrew Wilton and Tim Barringer, *American Sublime: Landscape Painting in the United States 1820–1880*.

43. Strong associations with romance and sexual passion established by nineteenth-century poetry, novels, and firsthand accounts drew honeymooners and wedding celebrations to Niagara Falls starting as early as the 1830s. See McKinsey, *Niagara Falls*. For historical photographs of Niagara Falls, see Daniel M. Dumych, *Images of America: Niagara Falls*. For analysis of Niagara Falls as a nineteenth-century tourist attraction, see J. F. Sears, *Sacred Places*. For an analysis of the industrial and recreational development around Niagara Falls, see Patrick Vincent McGreevy, *Imagining Niagara: The Meaning and Making of Niagara Falls*.

44. There are strong parallels between Hoover Dam and Niagara Falls. Both were sites of innovative developments in hydroelectric power, both became popular travel destinations and tourist attractions, and both have been firmly connected to the concept of the sublime (Hoover Dam as technological sublime and Niagara Falls as natural or American sublime). A thorough analysis of these parallels calls for its own study.

45. See the Bureau of Reclamation Website for an explanation of how the spillways system works: http://www.usbr.gov/lc/hooverdam/History/essays/spillways.html.

46. The spillways were used one subsequent time, July 2–September 6, 1983, as a result of a flood. Jet flow valve tests were conducted on April 4, June 14, September 11, and November 24, 1936; April 11–14 and December 9, 1938; February 27 and April 8, 1939; September 28, 1940; and August 6–18, September 12, and November 26, 1941. I am indebted to Colleen Dwyer, Public Affairs Specialist, US Bureau of Reclamation for this information.

47. This edition dedicated to "clarifying the historical and contemporary problems confronting Arizona" with regard to the Colorado River is highly critical of Arizona policies toward the dam. A full-page editorial cartoon on page 2 depicts Californians basking in water and electricity as they cash in on millions of dollars in federal government spending while laughing at Arizona politicians who continue to stubbornly refuse to ratify the Santa Fe Compact. Other articles in the issue are also critical of Arizona's policies regarding the dam and the river.

48. This image was used in other places as well; see, for example, *Arizona Highways*, July 1941.

49. In 1938 Cliff Segerblom accepted a position with the Bureau of Reclamation and became the first official photographer of the Boulder Canyon Project. His pictures were featured in *Life*, *Time*, and *National Geographic*, among other publications. For more information on Cliff Segerblom, see Dennis McBride, "Cliff Segerblom (1915–1990)."

50. Paul Ricoeur, *From Text to Action*, 168–69.

51. Paul Messaris, "What's Visual About 'Visual Rhetoric'?," 217.

52. Erik Cohen, "The Study of Touristic Images of Native People: Mitigating the Stereotype of a Stereotype," 49.

53. This image was also featured on the cover of Nye's *American Technological Sublime*, though no credit or citation of the photograph is provided in the book.

54. In fact, the picturesque, as an aesthetic theory, developed out of the popular idyllic landscape paintings of Claude Lorrain (1600-1682) and Nicolas Poussin (1594-1665), which signaled the beginnings of modern sightseeing wherein upper-class British "tourists" desiring to view the landscapes and places of classical antiquity for themselves embarked on the "Grand Tour" of Europe, a rite of passage for any aristocratic gentleman.

55. Malcolm Andrews, *Landscape and Western Art*, 16.

56. Judith Bernake, "Framing the View: Picturesque Landscape in Contemporary Tourism Imagery."

57. Ken Taylor, "Culture or Nature: Dilemmas of Interpretation," 74.

58. Bernake, "Framing the View."

59. David P. Billington and Donald C. Jackson, *Big Dams of the New Deal Era: A Confluence of Engineering and Politics*.

60. Jonathan Crary, *Suspensions of Perception: Attention, Spectacle, and Modern Culture*, 3.

5 / The Unseen

1. Linda M. Scott, "Images in Advertising: The Need for a Theory of Visual Rhetoric," 261. For a detailed discussion of this, see Gombrich, *Art and Illusion*.

2. For further discussion of how photography offers us knowledge about the past, see Philip J. Ethington, "Comment and Afterword: Photography and Placing the Past."

3. Azoulay discusses the significance of the photograph in tying spectatorship to a civic and moral duty toward the persons photographed. When a photograph is taken, Barthes suggests that it becomes a testimony that something "really was there." However, Azoulay complicates this notion by asserting that the "something" is often *people* who are caught up in the political act of photographing and being photographed, a space in which a civil negotiation takes place between governing and the subject's desire not to be governed. The governed are ruled through a complex relationship created by the state that is built largely on ideologies and thus is invisible to the citizenry, which allows the state to divide and mobilize the privileged citizenry against the unprivileged. Ariella Azoulay, *The Civil Contract of Photography*, 16–17.

4. Donald Worster, *Under Western Skies: Nature and History in the American West*, 69.

5. Ibid., 72.

6. Daniel Rosenberg, "No One Is Buried in Hoover Dam: Modernism and Rumor," 93.

7. Pat Lappin, Robert Ferraro, and Thelda Cox, Oral History of Winthrop A. Davis for the Boulder City Museum and Historical Association, n.d.

8. Winthrop A. Davis, "Unpublished Autobiography" (Boulder City Museum and Historical Association, n.d.), 27.

9. Lappin, Ferraro, and Cox, Oral History of Winthrop A. Davis.

10. W. A. Davis, "Unpublished Autobiography," 33.

11. Lappin, Ferraro, and Cox, Oral History of Winthrop A. Davis.

12. "The Lurking Power Behind Great Dams: California Disaster Focuses Our Attention on Other Great Structures," *New York Times*, March 18, 1928.

13. Ibid.

14. Catherine Mulholland, *William Mulholland and the Rise of Los Angeles*, 325.

15. According to Daniel Rosenberg, Hoover Dam's construction was "as much the result of carefully produced images as it is of carefully poured concrete." Rosenberg, "No One Is Buried in Hoover Dam," 93–94. For other similar skepticism regarding the permanence of dams, see Reisner, *Cadillac Desert*; Vilander, "Hoover Dam Photographs of Ben Glaha," 180–81; Donald Worster, "Hoover Dam: A Study in Domination."

16. In stark contrast, photographs taken by Six Companies photographers show burning trucks, wrecked cars, and damaged trailers. Some of these images by Six Companies Inc. photographer Walter J. Lubken, who was also the official photographer for the US Reclamation Service from 1903 to 1917, appeared in George Pettitt, *So Boulder Dam Was Built*. Although the book was sponsored and copyrighted by Six Companies Inc. and was intended as a promotional piece for the corporation, the narrative and the images in *So Boulder Dam Was Built* frankly depict the inherent dangers of working at the site, the same dangers that the Bureau of Reclamation worked to conceal. See Vilander, *Hoover Dam*, 46.

17. Lappin, Ferraro, and Cox, Oral History of Winthrop A. Davis.

18. Ibid.

19. Ibid.

20. "100 Percent Walk Out at Boulder Dam," *Industrial Worker*, August 15, 1931.

21. Lappin, Ferraro, and Cox, Oral History of Winthrop A. Davis.

22. "Evicts Strikers at Hoover Dam: Won't Arbitrate, Says Six Companies Head After Walkout," *Milwaukee Journal*, October 10, 1931.

23. For a detailed description of the strike, see Stevens, *Hoover Dam*, 69–78. For contemporaneous newspaper accounts, see the *Las Vegas Evening Review-Journal*, August 9, 13, 14, 15, 17, 24, 25, 1931; "1,400 Strikers Lose Hoover Dam Jobs," *New York Times*, August 9, 1931; "Strike of 150 Halts Work on Hoover Dam; Federal Officer Will Call Troops If Needed," *New York Times*, August 9, 1931; "Sheriff Prohibits Guns in Boulder Dam Strike Zone," *Fresno (CA) Bee*, August 9, 1931; "Strike Ties Up Work on Hoover Dam: Tunnel Workmen Walk as Wages Are Reduced," *Madison (WI) Capital Times*, August 9, 1931; "Strike Halts Construction on Boulder Dam," *Greely (CO) Tribune Republican*, August 10, 1931; "Hoover Dam Workmen Fight Discharge Move," *New York Times*, August 11, 1931; "Early Resumption on Dam Predicted," *New York Times*, August 12, 1931; "Ordered from Hoover Dam: Striking Workers, Facing Hunger, Mover Camp into Desert," *New York Times*, August 12, 1931; "Resume Work at Boulder Dam: District Placed Under Partial Martial Law After Short Strike," *Portsmouth (OR) Times*, August 14, 1931; "Boulder Dam Labor Protest Sent to Doak: Low Wages in Hell-Hole of Seat Stressed in Complaint," *Syracuse (NY) Herald*, August 19, 1931; "U.S. May Take Hand in Boulder Dam Labor Strike: Work at Standstill as Men Fail in Arbitration Move with Company," *Fresno (CA) Bee*, August 30, 1931.

24. F. L. Jones, "Spectacle of Beauty," 1946.

25. Stevens, *Hoover Dam*, 142–46.

26. Lappin, Ferraro, and Cox, Oral History of Winthrop A. Davis.

27. Brooks said of his photo, "I took the picture quickly, hid the camera under my coat and ducked into the crowd. A lot of people would have liked to wreck that picture." http://bytesdaily.blogspot.com/2012/04/pulitzer-prize-for-photography-1942 .html.

28. It is also important to consider that the physical dimensions of a newspaper in 1928 were considerably larger than what we are used to seeing today. With such a large canvas, the full-page display adds to the intensity of the illustration's effect on the viewer. The reproduction here simply cannot do justice to the size and impact of the original.

29. Rob Leicester Wagner, *Red Ink, White Lies: The Rise and Fall of Los Angeles Newspapers 1920–1962*.

30. Robert L. Duffus in "The Drama of the Colorado" refers to Southern California as a "premature Elysium" and laments the influence that newspaper "syndicates" such as the *Los Angeles Times* were having over the dam's construction and subsequent distribution of water. Elysium in Greek folklore was the place at the ends of the earth to which certain favored heroes were conveyed by the Greek gods after death. It is also a term that means a place or state of perfect happiness.

31. Contemporaneous perspectives on the dam project include F. L. Bird, "Who Will Benefit from Boulder Dam?"; Robert M. Brown, "The Utilization of the Colorado River"; Duffus, "Drama of the Colorado"; F. E. Weymouth, "Conservation of the Waters of the Colorado River from the Standpoint of the Reclamation Service." Volume 135 of the *Annals of the American Academy of Political and Social Science* catalogs several articles by prominent people in the project, such as Arthur P. Davis, Hiram W. Johnson, and E. O. Leatherwood, among others.

32. I would argue that the architecture in the image is art deco and not modern, moderne, or their derivatives (e.g., Streamline, Bauhaus, International, etc.), even though this edition of the paper was printed three years after the famous 1925 Paris *Exposition des Arts Decoratifs* (only posthumously named such in 1966), which featured modernist design styles that subsequently thrived in Europe as the dominant mode of architectural and design expression. Although art deco was passé in Europe after the early 1920s (even the art deco buildings constructed for the Paris exhibit were immediately razed thereafter), the art deco style remained popular in America well into the 1930s. See Alan Powers, "Art Deco: An Image Problem?"; Richard Striner, "Art Deco: Polemics and Synthesis"; Penelope Hunter, "Art Deco: The Last Hurrah."

33. For more on the artistry of art nouveau, see Paul Greenhalgh, *The Essence of Art Nouveau*.

34. James Grady, "Nature and the Art Nouveau"; Jiri Mucha, *Alphonse Mucha: Master of Art Nouveau*.

35. Interestingly, there are no vegetables depicted, only apples, oranges, pears, bananas, and grapes. The intent may have been to impart a more direct association with the "fruit of her (Nature's) bounty" or similar idiomatic or proverbial expressions.

36. Jan Thompson, "The Role of Women in the Iconography of Art Nouveau."

37. Elizabeth K. Menon, "The Functional Print in Commercial Culture: Henry Somm's Women in the Marketplace," 8.

38. Elizabeth Coffman, "Women in Motion: Loie Fuller and the 'Interpenetration' of Art and Science," 77.

39. Iconic women in flowing robes (e.g., the Statue of Liberty, Loie Fuller, and the female figure in the *Evening Herald*) meant to evoke "Americanicity," along with the visual iconography associated with female symbols representing the nation-state and nationalism ("Miss Liberty" or "Britannia"), are subjects that call for more investigation.

40. Margaret Haile Harris, *Loie Fuller: Magician of Light*.

41. Richard Current and Marcia Current, *Loie Fuller: Goddess of Light*.

42. One of the most widely circulated images publicizing the Imperial Valley, for example, called "California, the Cornucopia of the World," was originally prepared by the California Immigration Commission in 1885 but remained popular for many years thereafter. According to the poster, there is room in California for millions of immigrants. It also states that exactly "43,795,000 acres of government lands" remained "untaken," and additional railroad and private lands were available—enough for a million farmers. The poster promises that farmers will find in California a climate conducive to health and wealth without the threat of "cyclones or blizzards," a sentiment echoed by newspapers and other publications.

43. Grady, "Nature and the Art Nouveau," 191.

44. Robert Hariman, "Allegory and Democratic Public Culture in the Postmodern Era," 268.

45. Cäcilia Rentmeister, "Berufsverbot Für Die Musen. Warum Sind so Viele Allegorien Weiblich?"

46. Michelle Zimbalist Rosaldo, "Women, Culture, and Society: A Theoretical Overview," 21.

47. Ecofeminism combines ecocentric and feminist perspectives to critique patriarchal exploitation of land and animals (and women), valuing them only as commodities. See Marti Kheel, *Nature Ethics: An Ecofeminist Perspective*; Virginia J. Scharff, ed., *Seeing Nature Through Gender*; Ynestra King and Jael Miriam Silliman, eds., *Dangerous Intersections: Feminist Perspectives on Population, Environment, and Development*; Karen J. Warren, ed., *Ecofeminism: Women, Culture, Nature*; Haraway, *Simians, Cyborgs, and Women*.

48. Merchant, *Reinventing Eden*, 22. Ecofeminism also examines, among other things, gendered linguistic practices that reinforce the domination and oppression of women and the environment, using such terms as "raping" or "taming" the land, "nature's bounty," and "mother nature."

49. See also Merchant, *Reinventing Eden*, 10–62, in which she describes human history in the context of the Garden of Eden narrative and argues that since the Fall from the garden, humans have sought to recover or re-create Eden on earth (an earthly paradise).

50. Roderick Frazier Nash, *The Rights of Nature: A History of Environmental Ethics*, 144–45.

51. Lappin, Ferraro, and Cox, Oral History of Winthrop A. Davis.

52. D. L. Carmody, "What the Reclamation Women Do in Las Vegas."

53. For commentary on the role of race in labor issues of the early 1900s, see Jerry L. Simich and Thomas C. Wright, *The Peoples of Las Vegas: One City, Many Faces;* Tyler Stovall, "National Identity and Shifting Imperial Frontiers: Whiteness and the Exclusion of Colonial Labor After World War I"; William Sundstrom, "Down or Out? Unemployment and Occupational Shifts of Urban Black Men During the Great Depression"; Elmer R. Rusco, *Good Time Coming? Black Nevadans in the Nineteenth Century;* W. E. B. Du Bois, "The Negro in the American Social Order: Where Do We Go from Here?"

54. Quoted in Roosevelt Fitzgerald, "Blacks and the Boulder Dam Project," 258.

55. Roosevelt Fitzgerald, "The Evolution of a Black Community in Las Vegas, 1905–1940," 23.

56. Ibid., 24.

57. Ibid., 26–27.

58. Ibid., 25–26.

59. Fitzgerald, "Blacks and the Boulder Dam Project," 256.

60. Ibid., 256.

61. Ibid., 257

62. Ibid., 259.

63. Memorandum for the Secretary from Commissioner Meade, May 16, 1931.

64. For a detailed discussion of Glaha's work, see Vilander, *Hoover Dam,* 1999.

65. Vilander, "Hoover Dam Photographs of Ben Glaha," 184–86.

66. Robert V. Hine and John Mack Faragher, *The American West: A New Interpretive History.*

67. Worster, *Rivers of Empire,* 211.

68. In my research for this book, I identified a total of five photographs of Native Americans, although I am counting as one photograph a series of several images of Paiute depicted with Nevada governor James Scrugham since all were taken in relatively quick succession. There are also several additional illustrated depictions of Native Americans, but none of African Americans.

69. Interestingly, although a Bureau of Reclamation photographer took the black workers' photo, according to government records the Bureau of Indian Affairs is credited with these photos.

70. In fact, in 1871 the expedition, on the order of Wheeler, went up the Colorado River through Black Canyon.

71. I am indebted to Professor Bernadette Longo at the University of Minnesota, Twin Cities, for these insights.

72. Dorothy Childs Hogner, "Boulder Dam," 34.

6 / Imaging Labor

1. T. B. White, "Building the Big Dam."

2. Edmund Wilson, "Hoover Dam."

3. Quoted in Andrew J. Dunar and Dennis McBride, *Building Hoover Dam: An Oral History of The Great Depression*, 23, 68.

4. Stevens, *Hoover Dam*, 51.

5. Once at the site, most of the vehicles that families had used to get to the dam were abandoned, providing a business opportunity for one intrepid man who opened a junkyard full of broken-down cars and machinery.

6. We should keep in mind, however, that the transition depicted in these images is not one of pristine nature suddenly populated by humans; Hopi and Papago had already inhabited the area for thousands of years.

7. Quoted in Dunar and McBride, *Building Hoover Dam*, 40.

8. W. A. Davis, "Unpublished Autobiography," 29.

9. Stevens, *Hoover Dam*, 54.

10. W. A. Whynn, "White Man's Magic," unpublished memoir (Boulder City Museum and Historical Association, n.d.).

11. Ibid.

12. Winthrop A. Davis, "Letter to Mr. Wolf of the Nevada State Museum and Historical Society," June 18, 1990, Institutional Archives of Nevada State Museum, Las Vegas. Exhibits, 1991, Cahlan Research Library, Nevada State Museum, Las Vegas, NV.

13. W. A. Davis, "Unpublished Autobiography," 30.

14. There were two official Bureau of Reclamation photographers who documented the dam and several other photographers of note who published their work in various news and popular outlets. Ben Glaha (1899–1970) is the photographer responsible for the majority of the bureau's construction phase images. Glaha was hired by the Bureau of Reclamation in 1925 and was transferred to the bureau's Boulder Canyon Project's Division of Designs and Drafting in Boulder City, Nevada, in 1931. The bureau distributed his photographs to news organizations and publications around the world. While his images document in detail the dam's construction, they were also known for their aesthetic quality and composition. Glaha's photographs were the subject of commentary by Ansel Adams, appeared in photography books, and were exhibited in art galleries. Cliff Segerblom (1915–90) started as Hoover Dam photographer for the Bureau of Reclamation in 1938 and continued until 1941. Segerblom was tasked with photographing the project chiefly from a public relations standpoint. His photographs typically portray the dam as a recreation area and appear frequently in articles promoting tourism at the dam site. Other notable photographers are Louis J. Oakes (1877–1932), who sold his photographs of Las Vegas and the surrounding areas as postcards and framed prints throughout the 1920s, and Winthrop Davis (1904–2005), whose images centered on the squatter camps, labor strife, and poor living and working conditions of the people who worked on the dam project. Davis's photographs appeared in many media outlets, including the *Los Angeles Times*, *Salt Lake Tribune*, *Baltimore Sun*, International News Service, and *Pathé*.

15. E. Wilson, "Hoover Dam," 67.

16. Quoted in Dunar and McBride, *Building Hoover Dam*, 40.

17. W. A. Davis, "Unpublished Autobiography," 28.

18. E. Wilson, "Hoover Dam," 68.

19. W. A. Davis, "Unpublished Autobiography," 31.

20. Bruce Bliven, "The American Dnieperstroy."

21. See the US Bureau of Reclamation Website for a list of all fatalities at the dam: http://www.usbr.gov/lc/hooverdam/History/essays/fatal.html.

22. W. A. Davis, "Letter to Mr. Wolf."

23. Whynn, "White Man's Magic," 14.

24. Michael Hiltzik, *Colossus: Hoover Dam and the Making of the American Century*, 284–86.

25. Dennis McBride, e-mail message to author, August 24, 2012.

26. Rosenberg, "No One Is Buried in Hoover Dam."

27. "Plaque at Dam Is Made Issue in U.S. Senate," *Las Vegas Evening Review-Journal*, March 10, 1937.

28. Oskar J. W. Hansen, "From Bones of Water Pipe and Wool," 10.

29. Barbara Rigbey Connell, "Oskar Hansen, the Forgotten Artist of Hoover Dam," *Las Vegas Sun*, September 12, 1982.

30. Ibid.

31. See the Bureau of Reclamation's Website, "Artwork," *Bureau of Reclamation: Lower Colorado Region—Hoover Dam: Artwork*, September 10, 2001, accessed April 10, 2010, http://www.usbr.gov/lc/hooverdam/History/essays/artwork.html.

32. James M. Dennis, *Renegade Regionalists: The Modern Independence of Grant Wood, Thomas Hart Benton, and John Steuart Curry*.

33. Marlene Park and Gerald E. Markowitz, *Democratic Vistas: Post Offices and Public Art in the New Deal*.

34. Frances K. Pohl, *Framing America: A Social History of American Art*.

35. Park and Markowitz, *Democratic Vistas*, 68.

36. Painters such as Ben Shahn and Edward Hopper and photographers such as Lewis Hine, Alfred Stieglitz, Dorothea Lange, and Margaret Bourke-White are often considered social realist artists.

37. Erika Doss, "Looking at Labor: Images of Work in 1930s American Art."

38. Hine has been the subject of extensive scholarly research. For an overview of Hine in relation to American progressivism and documentary photography as social activism, see Alan Trachtenberg, *Reading American Photographs: Images as History, Mathew Brady to Walker Evans*, 164–230. For a detailed analysis of Hine's *Let Us Now Praise Famous Men*, see William Stott, *Documentary Expression and Thirties America*, 267–322. For a discussion of Hine's work for the Farm Security Administration (FSA) project, see Cara A. Finnegan, *Picturing Poverty: Print Culture and FSA Photographs*.

39. Kevin G. Barnhurst, "The Alternative Vision: Lewis Hine's 'Men at Work' and the Dominant Culture."

40. Trachtenberg, *Reading American Photographs*, 223–30.

41. Johnathan Weinberg, "I Want Muscle: Male Desire and the Image of the Worker in American Art of the 1930s," 119; Carl E. Van Horn, Herbert A. Schaffner, and Ray Marshall, eds., *Work in America: An Encyclopedia of History, Policy, and Society*, 634.

42. Barbara Melosh, "Manly Work: Public Art and Masculinity in Depression America."

43. Erika Doss, "Toward an Iconography of American Labor," 62–64.

44. The wpa Federal Art Project and the Treasury Section were two separate entities. Although the wpa is better known, the Treasury Section funded more than fourteen hundred works of art for federal buildings across the United States. For a discussion of the Treasury Section, see Melosh, "Manly Work."

45. Berger, *Ways of Seeing*, 54.

46. Ibid., 269.

47. Ibid., 9.

48. Barnhurst, "Alternative Vision," 87.

49. Ibid., 88.

50. Vilander, *Hoover Dam*, 69.

51. Barthes, *Camera Lucida*, 59.

52. Ibid., 14.

53. Gillian Dyer in *Advertising as Communication* presents a series of prompts for understanding how meaning is communicated in images of people.

54. Barnhurst, "Alternative Vision," 96.

55. Ibid., 85–107.

56. Anne Hollander, *Seeing Through Clothes*, 236.

57. Weinberg, "I Want Muscle," 115–34.

58. Hollander, *Seeing Through Clothes*, 209.

59. Mary Russo, *The Female Grotesque*, 8.

60. Stewart, *On Longing*, 105.

61. Quintilian, *Institutio Oratio*, XI, 3.85. For a detailed discussion of Cicero's and Quintilian's writings on and personal practice of hand gestures, see John Hall, "Cicero and Quintilian on the Oratorical Use of Hand Gestures."

62. Quintilian, *Institutio Oratio*, XI, 3.85–87.

63. Hogner, "Boulder Dam," 7.

64. Margret Borque-White photographed the Dnieper River dam in Dnieiperstoi, Russia—then the largest dam in the world—for *Life* magazine in 1930 before photographing Hoover Dam.

65. Lynn J. Rogers, "Mighty Work Impresses Tour Party: Will Require Three Years to Fill Reservoir Back of Structure," *Los Angeles Times*, December 9, 1934, sec. Outdoor.

66. R. G. Wilson, "Machine Age Iconography in the American West," 174, 193.

67. Susan Sontag, *On Photography*, 23.

68. Ibid., 4.

69. Stewart, *On Longing*, 75.

70. Gaston Bachelard, *The Poetics of Space*, 172–73.

71. Jeremy Bentham (1785) argued that his design of a "Panopticon" (pan = all; optic = seeing), structured in such a way that cells would be open to a central tower, would be a perfect prison because the prisoners would believe that they could be watched at any moment. Foucault writes, "The inmate must never know whether he is being

looked at any one moment; but he must be sure that he may always be so." Michel Foucault, *Discipline and Punish: The Birth of the Prison*, 201. See also Michel Foucault, "Panopticism," 356–67.

72. Foucault, *Discipline and Punish*, 205.

73. Amy E. Slaton, *Reinforced Concrete and the Modernization of American Building, 1900–1930*, 5.

74. Cecelia Tichi, *Shifting Gears: Technology, Literature, Culture in Modernist America*, 122.

75. Sharon Stockton, "Engineering Power: Hoover, Rand, Pound, and the Heroic Architect."

76. Tichi, *Shifting Gears*, 99.

77. Doss, "Looking at Labor," 232.

78. Ibid., 235–36.

7 / Advertising and Tourism

1. Berger, *Ways of Seeing*, 131–32.

2. For a discussion of the impact of stylistic elements in advertising that form visual rhetorical figures parallel to those found in language, see Edward F. McQuarrie and David Glen Mick, "Visual Rhetoric in Advertising: Text-Interpretive, Experimental, and Reader-Response Analyses." For a book-length study of the role of visual images in advertising, see Paul Messaris, *Visual Persuasion: The Role of Images in Advertising*. For a cultural study of advertising in early 1900s America, see Marchand, *Advertising the American Dream*.

3. Marguerite S. Shaffer, *See America First: Tourism and National Identity, 1880–1940*.

4. Sears, *Sacred Places*, 4.

5. Ibid., 5

6. McGreevy, *Imagining Niagara*, 2.

7. Nye, *American Technological Sublime*, 25.

8. Sears, *Sacred Places*, 3.

9. For a discussion of tourism and leisure, see Dean MacCannell, *The Tourist: A New Theory of the Leisure Class*.

10. John Urry, *The Tourist Gaze: Leisure and Travel in Contemporary Societies*, 1.

11. Cindy S. Aron, *Working at Play: A History of Vacations in the United States*, 252.

12. Frank Brunner, "Autocamping—the Fastest Growing Sport."

13. For the influence of the automobile on the national park system in America, see Hal K. Rothman, *Devil's Bargains: Tourism in the Twentieth-Century American West*, 143–67, in which he writes, "The automobile could reach such places, could negotiate narrow tracks and trails to places that railroad passengers could only dream of." For the influence of the automobile in shaping American national identity through tourism, see Shaffer, *See America First*, 130–68.

14. Aron, *Working at Play*, 253–54.

15. For commentary on various aspects of tourism in the American West, see David

M. Wrobel, Patrick T. Long, and Earl Pomeroy, eds., *Seeing and Being Seen: Tourism in the American West.*

16. Dyer, *Advertising as Communication*, 116.

17. Judith Williamson, *Decoding Advertisements.*

18. Barthes, "Rhetoric of the Image," 36.

19. John Berger writes, "When we 'see' a landscape, we situate ourselves in it. If we 'saw' the art of the past, we would situate ourselves in history. When we are prevented from seeing it, we are being deprived of the history which belongs to us"; Berger, *Ways of Seeing*, 11. Berger is writing about fine art, but the notion applies equally to the disappearing landscape caused by the dam and the rising waters of its lake.

Conclusion

1. Carole Blair and Neil Michel, "The Rushmore Effect: Ethos and National Collective Identity," 159.

2. Although the notion of dam removal has been more widely accepted, if not actively pursued, recently, it is certainly nothing new. In the American judicial system, for example, this concept has a long history. The Harvard Law Review, for instance, was publishing briefs in the early 1900s on the use of eminent domain to remove dams in the interest of public safety and health. One brief states, "It seems that the police power may be properly exercised to remove unhealthy conditions by cleaning out a non-navigable stream" (Harvard Law Review, 1912). Other notions of water access and dam policy have been debated for over a century, with the term "navigable waters" under dispute since the late 1700s and decisions such as *Arnold v Mundy* (1821) deciding that "navigable bays and rivers are common to all citizens" and *Illinois Central Railroad Company* v. *Illinois* (1892) marking the extent of the "public trust doctrine."

3. John Muir, *The Yosemite*, 257.

4. For a discussion of these two perspectives, see Keith Moxey, "Visual Studies and the Iconic Turn."

5. Ibid., 133.

6. Arjun Appadurai, "Introduction: Commodities and the Politics of Value."

Bibliography

Archives and Manuscript Collections

Archives of the Los Angeles Department of Water and Power
Bancroft Library at the University of California at Berkeley
Boulder City Museum and Historical Association
California State Library
County of Los Angeles Public Library
Doheny Memorial Library at the California Historical Society
Franklin D. Roosevelt Presidential Library and Museum
Herbert Hoover Presidential Library and Museum
Nevada State Museum
State Historical Society of Iowa
United States Bureau of Reclamation
United States National Archives
University of Nevada at Las Vegas Libraries, Department of Special Collections
University of Nevada at Reno
University of Southern California Libraries, Department of Special Collections
Wilson Library at the University of Minnesota

Other Sources

Adams, John R. *Damming the Colorado: The Rise of the Lower Colorado River Authority, 1933-1939.* College Station: Texas A&M University Press, 1979.

Andrews, Malcolm. *Landscape and Western Art.* New York: Oxford University Press, 1999.

Anreus, Alejandro, Diana L. Linden, and Jonathan Weinberg, eds. *The Social and the Real: Political Art of the 1930s in the Western Hemisphere.* University Park: Pennsylvania State University Press, 2006.

Appadurai, Arjun. "Introduction: Commodities and the Politics of Value." In *The Social Life of Things: Commodities in Cultural Perspective,* edited by Arjun Appadurai, 3-63. Cambridge: Cambridge University Press, 1988.

Arensberg, Mary. "Introduction: The American Sublime." In *The American Sublime,* edited by Mary Arensberg, 1-20. Albany: SUNY Press, 1986.

Armstrong, Carol. *Scenes in a Library: Reading the Photograph in the Book, 1843-1875.* Cambridge, MA: MIT Press, 1998.

Arnheim, Rudolf. *Art and Visual Perception: A Psychology of the Creative Eye.* Berkeley: University of California Press, 2004.

———. *Visual Thinking.* Berkeley: University of California Press, 2004.

Aron, Cindy S. *Working at Play: A History of Vacations in the United States.* New York: Oxford University Press, 2001.

Arwas, Victor, Jana Brabcova-Orlikova, and Anna Dvorak. *Alphonse Mucha: The Spirit of Art Nouveau.* New Haven, CT: Yale University Press, 1998.

Asen, Robert. "Women, Work, Welfare: A Rhetorical History of Images of Poor Women in Welfare Policy Debates." *Rhetoric & Public Affairs* 6 (2003): 285–312.

Azoulay, Ariella. *The Civil Contract of Photography.* New York: Zone Books, 2008.

Bachelard, Gaston. *The Poetics of Space.* Translated by Maria Jolas. Boston: Beacon Press, 1994.

Banta, Martha. *Imaging American Women: Idea and Ideals in Cultural History.* New York: Columbia University Press, 1987.

Barnhurst, Kevin G. "The Alternative Vision: Lewis Hine's 'Men at Work' and the Dominant Culture." In *Photo-Textualities: Reading Photographs and Literature,* edited by Marsha Bryant, 85–107. Newark: University of Delaware Press, 1996.

Barthes, Roland. *Camera Lucida: Reflections on Photography.* New York: Hill and Wang, 1982.

———. *Mythologies.* New York: Hill and Wang, 1972.

———. "The Photographic Message." In *Image Music Text,* edited and translated by Stephen Heath, 15–31. London: Fontana Press, 1977.

———. "The Rhetoric of the Image." In *Image Music Text,* edited and translated by Stephen Heath, 32–51. London: Fontana Press, 1977.

Barton, Ben F., and Marthalee S. Barton. "Ideology and the Map: Toward a Postmodern Visual Design Practice." In *Central Works in Technical Communication,* edited by Johndan Johnson-Eilola and Stuart Selber, 232–54. New York: Oxford University Press, 2004.

Baudrillard, Jean. *America.* Translated by Chris Turner. London: Verso, 1989.

Bayer, Patricia. *Art Deco Architecture: Design, Decoration, and Detail from the Twenties and Thirties.* London: Thames and Hudson, 1999.

Beckley, Bill. "Sticky Sublime." In *Sticky Sublime,* edited by Bill Beckley, 2–15. New York: Allworth Press, 2001.

Bell, Claudia, and John Lyall. *The Accelerated Sublime: Landscape, Tourism, and Identity.* Westport, CT: Praeger, 2001.

Bellush, Bernard. *Franklin D. Roosevelt as Governor of New York.* New York: Columbia University Press, 1955.

Benjamin, Walter. *The Arcades Project.* Edited by Rolf Tiedemann. Translated by Howard Eiland and Kevin McLaughlin. Cambridge, MA: Harvard University, Belknap Press, 2002.

———. *Illuminations: Essays and Reflections.* New York: Schocken, 1969.

———. *One-Way Street and Other Writings.* London: Verso, 1997.

Berger, John. *About Looking.* First Vintage International ed. New York: Vintage, 1992.

———. *Ways of Seeing.* London: Penguin, 1972.

Bernake, Judith. "Framing the View: Picturesque Landscape in Contemporary Tourism Imagery." Paper presented at the Hawaii International Conference on Arts and Humanities, January 2004.

Bernstein, Michael A. *The Great Depression: Delayed Recovery and Economic Change in America, 1929–1939.* Cambridge: Cambridge University Press, 1989.

Billington, David P., and Donald C. Jackson. *Big Dams of the New Deal Era: A Confluence of Engineering and Politics.* Norman: University of Oklahoma Press, 2006.

Bird, F. L. "Who Will Benefit from Boulder Dam?" *New Republic,* July 30, 1930, 310–13.

Bird, S. Elizabeth, ed. *Dressing in Feathers: The Construction of the Indian in American Popular Culture.* Boulder, CO: Westview Press, 1996.

Blair, Carole. "Contemporary U.S. Memorial Sites as Exemplars of Rhetoric's Materiality." In *Rhetorical Bodies,* edited by Jack Selzer and Sharon Crowley. Madison: University of Wisconsin Press, 1999.

Blair, Carole, George Dickinson, and Brian L. Ott, eds. *Places of Public Memory: The Rhetoric of Museums and Memorials.* Tuscaloosa: University of Alabama Press, 2010.

Blair, Carole, Marsha S. Jeppeson, and Enrico Pucci Jr. "Public Memorializing in Postmodernity: The Vietnam Veterans Memorial as Prototype." *Quarterly Journal of Speech* 77 (1991): 263–88.

Blair, Carole, and Neil Michel. "The Rushmore Effect: Ethos and National Collective Identity." In *The Ethos of Rhetoric,* edited by Michael J. Hyde, 156–96. Columbia: University of South Carolina Press, 2004.

Bliven, Bruce. "The American Dnieperstroy." *New Republic,* December 11, 1935, 125–27.

Bloom, Harold. "Emerson and Whitman: The American Sublime." In *Sublime,* edited by Bill Beckley, 16–39. New York: Allworth Press, 2001.

Bond, J. Max. "The Educational Programs for Negroes in the TVA." *Journal of Negro Education* 6 (1937): 144–51.

———. "The Training Program of the Tennessee Valley Authority for Negroes." *Journal of Negro Education* 7 (1938): 383–89.

"Boulder Dam—Lake Mead." *Nevada Magazine,* August 1945, 16–17.

Braude, Lee. *Work and Workers: A Sociological Analysis.* New York: Praeger, 1975.

Breeze, Carla. *American Art Deco: Modernistic Architecture and Regionalism.* New York: W. W. Norton, 2003.

Brown, Robert M. "The Utilization of the Colorado River." *Geographical Review* 17 (1927): 453–66.

Brunner, Frank. "Autocamping--the Fastest Growing Sport." *Outlook,* July 16, 1924, 437–39.

Bryson, Norman, Michael Ann Holly, and Keith Moxey, eds. *Visual Culture: Images and Interpretations.* Middletown, CT: Wesleyan University Press, 1994.

Buell, Lawrence. *The Environmental Imagination: Thoreau, Nature Writing, and the Formation of American Culture.* Cambridge, MA: Harvard University, 1996.

———. *The Future of Environmental Criticism: Environmental Crisis and Literary Imagination.* Hoboken, NJ: Wiley-Blackwell, 2005.

———. "Letter." *PMLA Forum on Literatures of the Environment* 114.5 (1999): 1090–92.

Burgin, Victor. *In/Different Spaces: Place and Memory in Visual Culture.* Berkeley: University of California Press, 1996.

Burke, Kenneth. *The Philosophy of Literary Form.* 3rd ed. Berkeley: University of California Press, 1974.

———. *A Rhetoric of Motives.* Berkeley: University of California Press, 1969.

Campbell, Robert B. "Newlands, Old Lands: Native American Labor, Agrarian

Ideology, and the Progressive-Era State in the Making of the Newlands Reclamation Project, 1902–1926." *Pacific Historical Review* 71 (2002): 203–38.

Cannon, Brian Q. "Power Relations: Western Rural Electric Cooperatives and the New Deal." *Western Historical Quarterly* 31 (2000): 133–60.

Cantrill, James G., and Christine L. Oravec, eds. *The Symbolic Earth: Discourse and Our Creation of the Environment.* Lexington: University Press of Kentucky, 1996.

Carion, Anne. "Alfons Mucha (1860–1929)." *FMR* 15 (1995): 8–9.

Carmody, D. L. "What the Reclamation Women Do in Las Vegas." *New Reclamation Era* 22 (August 1931): 202–3.

Certeau, Michel de. *The Practice of Everyday Life.* Translated by Steven Rendall. Berkeley: University of California Press, 1984.

Chan, Loren B. "The Chinese in Nevada: An Historical Survey, 1856–1970." In *Chinese on the American Frontier,* edited by Arif Dirlik and Malcolm Yeung, 85–116. New York: Rowman and Littlefield, 2003.

Chase, John. "The Role of Consumerism in American Architecture." *Journal of Architectural Education (1984–)* 44 (1991): 211–24.

Clark, Gregory. *Rhetorical Landscapes in America: Variations on a Theme from Kenneth Burke.* Illustrated ed. Columbia: University of South Carolina Press, 2004.

Clements, Kendrick A. "Herbert Hoover and Conservation 1921–33." *American Historical Review* 89 (1984): 67–88.

Coffman, Elizabeth. "Women in Motion: Loie Fuller and the 'Interpenetration' of Art and Science." *Camera Obscura* 17 (2002): 72–105.

Cohen, Erik. "The Study of Touristic Images of Native People: Mitigating the Stereotype of a Stereotype." In *Tourism Research: Critiques and Challenges,* edited by Douglas G. Pearce and Richard Warren Butler, 36–69. New York: Routledge, 1993.

Cole, Olen. "The Negro in the TVA." *Opportunity* 12 (1934): 111–12.

Contreras, Belisario R. *Tradition and Innovation in New Deal Art.* Lewisburg, PA: Bucknell University Press, 1984.

Cooper, Emmanuel. *The Sexual Perspective: Homosexuality and Art in the Last 100 Years in the West.* London: Routledge and Kegan Paul, 1986.

Coray, Michael S. "'Democracy' on the Frontier: A Case Study of Nevada Editorial Attitudes on the Issue of Nonwhite Equality." *Nevada Historical Society Quarterly* 21 (1978): 189–204.

Corbett, Julia B. *Communicating Nature: How We Create and Understand Environmental Messages.* Washington, DC: Island Press, 2006.

Corwin, Sharon. "Picturing Efficiency: Precisionism, Scientific Management, and the Effacement of Labor." *Representations* 84 (2003): 139–65.

Cox, Thomas R. "Before the Casino: James G. Scrugman, State Parks, and Nevada's Quest for Tourism." *Western Historical Review* 24 (1993): 332–50.

Crary, Jonathan. *Suspensions of Perception: Attention, Spectacle, and Modern Culture.* Cambridge, MA: MIT Press, 2001.

———. *Techniques of the Observer: On Vision and Modernity in the 19th Century.* Cambridge, MA: MIT Press, 1992.

Cronon, William. *Changes in the Land: Indians, Colonists, and the Ecology of New England.* Revised ed. New York: Hill and Wang, 2003.

———. *Nature's Metropolis: Chicago and the Great West.* New York: W. W. Norton, 1991.

———. "Revisiting the Vanishing Frontier: The Legacy of Frederick Jackson Turner." *Western Historical Quarterly* 18 (1987): 157–76.

———. "The Trouble with Wilderness; or, Getting Back to the Wrong Nature." *Guernica / A Magazine of Art & Politics,* May 1, 2005, http://www.guernicamag.com /features/41/the_trouble_with_wilderness_or/.

———, ed. *Uncommon Ground: Rethinking the Human Place in Nature.* New ed. New York: W. W. Norton, 1996.

Cross, Whitney R. "Ideas in Politics: The Conservation Policies of the Two Roosevelts." *Journal of the History of Ideas* 14 (1953): 421–38.

Crouch, David, and Nina Lubbren. *Visual Culture and Tourism.* English ed. Oxford: Berg, 2003.

Current, Richard, and Marcia Current. *Loie Fuller: Goddess of Light.* Boston: Northeastern University Press, 1997.

Curtis, William. *Modern Architecture Since 1900.* 3rd ed. New York: Paidon Press, 1996.

Cutler, Phoebe. *The Public Landscape of the New Deal.* New Haven, CT: Yale University Press, 1986.

Dabkis, Melissa. *Visualizing Labor in American Sculpture: Monuments, Manliness, and the Work Ethic, 1880–1935.* New York: Cambridge University Press, 1999.

Danly, Susan, and Leo Marx, eds. *Railroad in American Art: Representations of Technological Change.* Cambridge, MA: MIT Press, 1988.

Davey, Gerald. "A Historical Approach to Understanding Documentary Photographs: Dialogue, Interpretation, and Method." In *Handbook of Visual Communication: Theory, Methods, and Media,* edited by Kenneth L. Smith, Sandra Moriarty, Gretchen Barbatsis, and Keith Kenny, 565–76. Mahwah, NJ: Lawrence Erlbaum, 2004.

Davis, Arthur P. "Problems of the Colorado River, Relative Advantages of the Boulder Site." *Annals of the American Academy of Political and Social Science* 135 (1928): 123–26.

———. "What the Boulder Dam Project Means to California and to the Nation." *Annals of the American Academy of Political and Social Science* 135 (1928): 127–32.

Davis, John P. "The Plight of the Negro in the TVA." *Crisis* 42 (1935): 294–95, 314–15.

Davis, Susan G. "Landscapes of Imagination: Tourism in Southern California." *Pacific Historical Review* 68 (1999): 173–91.

Deluca, Kevin M., and Anne T. Demo. "Imaging Nature: Watkins, Yosemite, and the Birth of Environmentalism." *Critical Studies in Mass Communication* 17 (2000): 241–61.

Dennis, James M. *Renegade Regionalists: The Modern Independence of Grant Wood, Thomas Hart Benton, and John Steuart Curry.* Madison: University of Wisconsin Press, 1998.

Derrida, Jacques. *The Truth in Painting.* Chicago: University of Chicago Press, 1987.

Dikovitskaya, Margaret. *Visual Culture: The Study of the Visual After the Cultural Turn.* Cambridge, MA: MIT Press, 2005.

Dodd, Douglass W. "Boulder Dam Recreation Area: The Bureau of Reclamation, the National Park Service, and the Origins of the National Recreation Area Concept at Lake Mead, 1929–1936." In *The Bureau of Reclamation: History Essays from the Centennial Symposium Volumes I and II*, 467–94. Denver, CO: Bureau of Reclamation, US Department of the Interior, 2008.

Doherty, Elizabeth M. "Viewing Work Historically Through Art: Incorporating the Visual Arts into Organizational Studies." *Journal of Management History* 12 (2006): 137–53.

Dolen, Timothy P. "Historical Development of Durable Concrete for the Bureau of Reclamation." In *The Bureau of Reclamation: History Essays from the Centennial Symposium Volumes I and II*, 135–51. Denver, CO: Bureau of Reclamation, US Department of the Interior, 2008.

Doss, Erika. "Looking at Labor: Images of Work in 1930s American Art." *Journal of Decorative and Propaganda Arts* 24 (2002): 231–57.

———. "Toward an Iconography of American Labor: Work, Workers, and the Work Ethic in American Art 1930–1945." *Design Issues* 15 (1997): 53–66.

Du Bois, W. E. B. "The Negro in the American Social Order: Where Do We Go from Here?" *Journal of Negro Education* 8 (1939): 551–70.

Duffus, Robert L. "The Drama of the Colorado." *New Republic*, April 1, 1925, 147–49.

Dumych, Daniel M. *Images of America: Niagara Falls*. Mount Pleasant, SC: Arcadia, 1996.

Dunar, Andrew J., and Dennis McBride. *Building Hoover Dam: An Oral History of the Great Depression*. Reno: University of Nevada Press, 1993.

Dunaway, Finis. *Natural Visions: The Power of Images in American Environmental Reform*. Chicago: University of Chicago Press, 2008.

Duncan, Alastair. "Art Deco Lighting." *Journal of Decorative and Propaganda Arts* 1 (1986): 20–31.

Durden, Mark. "The Limits of Modernism: Walker Evans and James Agee's 'Let Us Now Praise Famous Men.'" In *Literary Modernism and Photography*, edited by Paul Hansom, 25–30. Westport, CT: Praeger, 2002.

Dyer, Gillian. *Advertising as Communication*. Reprint ed. London: Routledge, 1982.

Eagleton, Terry. *The Idea of Culture*. Wiley-Blackwell, 2000.

———. *Ideology: An Introduction*. New ed. London: Verso, 2007.

Edwards, Janis L. "Echoes of Camelot: How Images Construct Cultural Memory Through Rhetorical Framing." In *Defining Visual Rhetorics*, edited by Charles A. Hill and Marguerite Helmers. Madison: University of Wisconsin Press, 2004.

Edwards, Janis L., and Carol K. Winkler. "Representative Form and the Visual Ideograph: The Iwo Jima Image in Editorial Cartoons." *Quarterly Journal of Speech* 83 (1997): 289–310.

Ekirch, Arthur Alphonse. *Ideologies and Utopias: The Impact of the New Deal on American Thought*. Chicago: Times Books, 1969.

Elliott, Russell R., and William D. Rowley. *History of Nevada*. 2nd ed. Lincoln: University of Nebraska Press, 1987.

Ethington, Philip J. "Comment and Afterword: Photography and Placing the Past." *Journal of Visual Culture* 9 (2010): 439–48.

Evans, Jessica, and Stuart Hall, eds. *Visual Culture: The Reader.* London: Sage, 1999.

Evernden, Neil. *The Social Creation of Nature.* Baltimore: Johns Hopkins University Press, 1992.

Finnegan, Cara A. "The Naturalistic Enthymeme and Visual Argument: Photographic Representation in the 'Skull Controversy.'" *Argumentation & Advocacy* 37 (2001): 133–49.

———. *Picturing Poverty: Print Culture and FSA Photographs.* Washington, DC: Smithsonian, 2003.

———. "Recognizing Lincoln: Image Vernaculars in Nineteenth-Century Visual Culture." *Rhetoric & Public Affairs* 8 (2005): 31–58.

———. "Review Essay: Visual Studies and Visual Rhetoric." *Quarterly Journal of Speech* 90 (2004): 234–56.

———. "What Is This a Picture Of? Some Thoughts on Images and Archives." *Rhetoric & Public Affairs* 9 (2006): 116–23.

Finnegan, Cara A., and Jennifer L. Jones-Barbour. "Review Essay: Visualizing Public Address." *Rhetoric & Public Affairs* 9.3 (2006): 489–532.

Fitzgerald, Roosevelt. "Blacks and the Boulder Dam Project." *Nevada Historical Society Quarterly* 24 (1981): 255–61.

———. "The Evolution of a Black Community in Las Vegas, 1905–1940." *Nevada Public Affairs Review* 2 (1987): 23–28.

Fleckenstein, Kristie S., Sue Hum, and Linda T. Calendrillo, eds. *Ways of Seeing, Ways of Speaking: The Integration of Rhetoric and Vision in Constructing the Real.* West Lafayette, IN: Parlor Press, 2007.

Foner, Philip S., and Ronald L. Lewis, eds. *The Black Workers: A Documentary History from Colonial Times to the Present.* Vol. 6, *The Era of Post-war Prosperity and the Great Depression, 1920–1936.* Philadelphia: Temple University Press, 1981.

Foss, Sonja K. "Review Essay: Visual Imagery as Communication." *Text and Performance Quarterly* 12 (1992): 85–96.

———. "The Rhetoric of Visual Arguments." In *Defining Visual Rhetorics,* edited by Charles A. Hill and Marguerite Helmers, 303–14. Mahwah, NJ: Lawrence Erlbaum, 2004.

Foucault, Michel. *Discipline and Punish: The Birth of the Prison.* Translated by Alan Sheridan. New York: Vintage, 1995.

———. "Of Other Spaces: Utopias and Heterotopias." In *Rethinking Architecture: A Reader in Cultural Theory,* edited by Neil Leach, 350–56. New York: Routledge, 1997.

———. "Panopticism." In *Rethinking Architecture: A Reader in Cultural Theory,* edited by Neil Leach, 356–67. New York: Routledge, 1997.

Fradkin, Philip L. *A River No More: The Colorado River and the West.* Berkeley: University of California Press, 1996.

Gallagher, Victoria, and Kenneth S. Zagacki. "Visibility and Rhetoric: The Power of Visual Images in Norman Rockwell's Depictions of Civil Rights." *Quarterly Journal of Speech* 91 (2005): 175.

Garrard, Greg. *Ecocriticism*. New York: Routledge, 2004.

Gartner, William C. *Tourism Development: Principles, Processes, and Policies*. New York: John Wiley and Sons, 1996.

Gillman, Howard. *The Constitution Besieged: The Rise and Demise of Lochner Era Police Powers Jurisprudence* (Durham, NC: Duke University Press, 1995).

Goankar, Dilip. "Toward New Imaginaries: An Introduction." *Public Culture* 14 (2002): 1–19.

Goetzmann, William H. *Exploration and Empire: The Explorer and the Scientist in the Winning of the American West*. New York: Alfred P. Knopf, 1966.

Gombrich, E. H. *Art and Illusion: A Study in the Psychology of Pictorial Representation*. 6th ed. London: Phaidon Press, 2004.

Gordon, Colin. *New Deals: Business, Labor, and Politics in America, 1920–1935*. Cambridge: Cambridge University Press, 1994.

Grady, James. "Nature and the Art Nouveau." *Art Bulletin* 37 (1955): 187–92.

Grant, Nancy L. *TVA and Black Americans: Planning for the Status Quo*. Philadelphia: Temple University Press, 1990.

Green, Nancy L. "The Comparative Gaze: Travelers in France Before the Era of Mass Tourism." *French Historical Studies* 25 (2002): 423–40.

Green, Nicholas. *The Spectacle of Nature: Landscape and Bourgeois Culture in Nineteenth-Century France*. Manchester, UK: Manchester University Press, 1993.

Greenhalgh, Paul. *The Essence of Art Nouveau*. New York: Harry N. Abrams, 2000.

———. "'A Great Seriousness': Art Nouveau and the Status of Style." *Apollo* 76 (1962): 3–10.

Grove, Richard H. *Green Imperialism: Colonial Expansion, Tropical Island Edens and the Origins of Environmentalism, 1600–1860*. Cambridge: Cambridge University Press, 1995.

Gruening, Ernest. "Power and Propaganda." *American Economic Review, Supplement, Papers, Proceedings of the Forty-third Annual Meeting of the American Economic Association* 21 (1931): 202–41.

Gunn, Joshua, and David Beard. "On the Apocalyptic Sublime." *Southern Speech Communication Journal* 65 (2000): 269–86.

Halbwachs, Maurice. *On Collective Memory*. Edited and translated by Lewis A. Coser. Chicago: University of Chicago Press, 1992.

Hall, Dennis R., ed. "Niagara Falls." In *American Icons [Three Volumes]: An Encyclopedia of the People, Places, and Things That Have Shaped Our Culture*, 493–500. Santa Barbara, CA: Greenwood Press, 2006.

Hall, John. "Cicero and Quintilian on the Oratorical Use of Hand Gestures." *Classical Quarterly* 54 (2004): 143–60.

Hall, Stuart. "The Determinations of News Photographs." In *Manufacture of the News: Deviance Social Problems and the Mass Media*, edited by Stanley Cohen and Jock Young, 176–90. London: Constable, 1973.

———. *Representation: Cultural Representations and Signifying Practices*. London: Sage, 1997.

Hall, Thomas E., and J. David Ferguson. *The Great Depression: An International Disaster of Perverse Economic Policies*. Ann Arbor: University of Michigan Press, 1998.

Hammond, Scott C., Rob Anderson, and Kenneth N. Cissna. "The Problematics of Dialogue and Power." In *Communication Yearbook 27*, edited by Pamela J. Kalbfleisch, 125–53. Mahwah, NJ: Lawrence Erlbaum, 2003.

Hand, Samuel B. "Roseman, Thucydides, and the New Deal." *Journal of American History* 55 (1986): 334–48.

Hansen, Oskar J. W. "From Bones of Water Pipe and Wool." *The Reclamation Era*, February 1942, 8–10.

———. "A Split Second Petrified on the Face of the Universal Clock." *The Reclamation Era*, February 1942, 5–7.

———. "With the Look of Eagles." *The Reclamation Era*, February 1942, 2–4.

Haraway, Donna J. "The Persistence of Vision." In *The Visual Culture Reader*, edited by Nicholas Mirzoeff, 191–98. London: Routledge, 1998.

———. *Simians, Cyborgs, and Women: The Reinvention of Nature*. New York: Routledge, 1991.

Hariman, Robert. "Allegory and Democratic Public Culture in the Postmodern Era." *Philosophy and Rhetoric* 35 (2002): 267–96.

———. "For the Sake of Argument: Practical Reasoning, Character, and the Ethics of Belief." *Quarterly Journal of Speech* 91 (2005): 456–58.

Hariman, Robert, and John Louis Lucaites. *No Caption Needed: Iconic Photographs, Public Culture, and Liberal Democracy*. Chicago: University of Chicago Press, 2007.

———. "Performing Civic Identity: The Iconic Photograph of the Flag Raising at Iwo Jima." *Quarterly Journal of Speech* 88 (2002): 363–92.

Harley, J. B. "Deconstructing the Map." In *Postmodernism: Critical Concepts*, edited by Victor E. Taylor and Charles E. Winquist, 3–24. New York: Routledge, 2000.

Harris, Jonathan. *Federal Art and National Culture: The Politics of Identity in New Deal America*. Cambridge: Cambridge University Press, 1995.

Harris, Margaret Haile. *Loie Fuller: Magician of Light*. Richmond: Virginia Museum of Art, 1979.

Hattenhauer, Darryl. "The Rhetoric of Architecture: A Semiotic Approach." *Communication Quarterly* 31 (1984): 71–77.

Hawkes, David. *Ideology*. 2nd ed. New York: Routledge, 2003.

Hembold, Louis Rita. "Downward Occupational Mobility During the Great Depression: Urban Black and White Working Class Women." *Labor History* 29 (1988): 135–72.

Herndl, Carl. *Green Culture: Environmental Rhetoric in Contemporary America*. Madison: University of Wisconsin Press, 1996.

Hetherington, Kevin. *Badlands of Modernity: Heterotopia and Social Ordering*. New York: Routledge, 1997.

Hiltzik, Michael. *Colossus: Hoover Dam and the Making of the American Century*. Free Press, 2010.

Hine, Lewis Wickes, and International Museum of Photography at George Eastman House. *Men at Work*. Mineola, NY: Dover, 1977.

Hine, Robert V., and John Mack Faragher. *The American West: A New Interpretive History.* New Haven, CT: Yale University Press, 2000.

Hogner, Dorothy Childs. "Boulder Dam." *American Girl: The Magazine for All Girls Published by the Girl Scouts,* June 1941, 5–7, 30, 33–34.

Hollander, Anne. *Seeing Through Clothes.* New York: Viking, 1978.

Holly, Michael Ann. *Past Looking: Historical Imagination and the Rhetoric of the Image.* Ithaca, NY: Cornell University Press, 1996.

Horkheimer, Max, and Theodor W. Adorno. *Dialectic of Enlightenment.* Stanford: Stanford University Press, 2007.

Houck, Davis W. *Rhetoric as Currency: Hoover, Roosevelt, and the Great Depression.* College Station: Texas A&M University Press, 2001.

Hughes, Thomas P. *American Genesis: A Century of Invention and Technological Enthusiasm, 1870–1970.* Chicago: University of Chicago Press, 2004.

———. *Human-Built World: How to Think About Technology and Culture.* Chicago: University of Chicago Press, 2004.

Huhndorf, Shari M. *Going Native: Indians in the American Cultural Imagination.* Ithaca, NY: Cornell University Press, 2001.

Hundley, Norris, Jr. *The Great Thirst: Californians and Water—a History.* Revised ed. Berkeley: University of California Press, 2001.

———. *Water and the West: The Colorado River Compact and the Politics of Water in the American West.* 2nd ed. Berkeley: University of California Press, 2009.

Hunnicutt, Benjamin Kline. *Work Without End: Abandoning Shorter Hours for the Right to Work.* Philadelphia: Temple University Press, 1988.

Hunter, Jefferson. *Image and Word: The Interaction of Twentieth-Century Photographs and Texts.* Cambridge, MA: Harvard University Press, 1987.

Hunter, Penelope. "Art Deco: The Last Hurrah." *Metropolitan Museum of Art Bulletin* 30 (1972): 257–67.

Jackson, Donald C. *Building the Ultimate Dam: John S. Eastwood and the Control of Water in the West.* Lawrence: University Press of Kansas, 1995.

———, ed. *Dams (Studies in the History of Civil Engineering, V. 4).* Surrey, UK: Ashgate, 1998.

———. "Origins of Boulder/Hoover Dam: Siting, Design, and Hydroelectric Power." In *The Bureau of Reclamation: History Essays from the Centennial Symposium Volumes I and II,* 273–88. Denver, CO: Bureau of Reclamation, US Department of the Interior, 2008.

Jay, Martin. "Cultural Relativism and the Visual Turn." *Journal of Visual Culture* 1 (2002): 267–79.

———. *Downcast Eyes: The Denigration of Vision in Twentieth-Century French Thought.* Berkeley: University of California Press, 1994.

Jefferson, Thomas. "Notes on the State of Virginia: Query XIX." In *Thomas Jefferson: Writings: Autobiography, Notes on the State of Virginia, Public and Private Papers, Addresses, Letters,* edited by Merrill D. Peterson, 290. New York: Library of America, 1984.

———. *The Papers of Thomas Jefferson.* Vol. 1, 1760–1776. Edited by Julian P. Boyd. Princeton, NJ: Princeton University Press, 1950.

———. "To George Washington, 1787." In *The Writings of Thomas Jefferson: Memorial Edition*, edited by Andrew Adgate Lipscomb and Albert Ellery Bergh, 6:277. Washington, DC, 1903.

———. "To John Jay, 1785." In *Thomas Jefferson: Writings: Autobiography, Notes on the State of Virginia, Public and Private Papers, Addresses, Letters*, edited by Merrill D. Peterson, 818–20. New York: Library of America, 1984.

Jenks, Chris. *Visual Culture.* New York: Routledge, 1995.

Johnson, Hiram W. "The Boulder Canyon Project." *Annals of the American Academy of Political and Social Science* 135 (February 1928): 150–56.

Jones, Gareth Stedman. *Outcast London.* New York: Pantheon, 1984.

Josephson, Paul R. *Industrialized Nature: Brute Force and the Transformation of the Natural World.* Washington, DC: Island Press, 2002.

Kahrl, William L. *Water and Power: The Conflict over Los Angeles Water Supply in the Owens Valley.* Berkeley: University of California Press, 1983.

Kant, Immanuel. *Critique of Judgment.* Edited by Nicholas Walker. Translated by James Creed Meredith and Nicholas Walker. New York: Oxford University Press, 2007.

———. *Observations on the Feeling of the Beautiful and Sublime.* Translated by John T. Golthwait. 2nd ed. Berkeley: University of California Press, 2004.

Kasson, John F. *Civilizing the Machine: Technology and Republican Values in America, 1776–1900.* New York: Hill and Wang, 1976.

Kaufmann, Gordon B. "The Architecture of Boulder Dam." *Architectural Concrete* 2 (1936): 2–5.

Kennedy, David M. *Freedom from Fear: The American People in Depression and War, 1929–1945.* New York: Oxford University Press, 2001.

Kennedy, Roger G., and David Larkin. *When Art Worked: The New Deal, Art, and Democracy.* New York: Rizzoli, 2009.

Kenney, Keith. "Building Visual Communication Theory by Borrowing from Rhetoric." In *Visual Rhetoric in a Digital World: A Critical Sourcebook*, edited by Carolyn Handa, 321–43. New York: Bedford/St. Martin's, 2004.

Kershner, Frederick D., Jr. "George Chaffey and the Irrigation Frontier." *Agricultural History* 27 (October 1953): 115–22.

Kheel, Marti. *Nature Ethics: An Ecofeminist Perspective.* Lanham, MD: Rowman and Littlefield, 2007.

King, Ynestra, and Jael Miriam Silliman, eds. *Dangerous Intersections: Feminist Perspectives on Population, Environment, and Development.* Cambridge, MA: South End Press, 1999.

Kirby, John B. *Black American Roosevelt Era: Liberalism and Race.* Knoxville: University of Tennessee Press, 1982.

Koerner, J. L. "Borrowed Sight: The Halted Traveller in Caspar David Friedrich and William Wordsworth." *Word & Image* 1 (1985): 149–63.

Krall, Lisa. "Thomas Jefferson's Agrarian Vision and the Changing Nature of Property." *Journal of Economic Issues* 36 (2002): 131–50.

Kress, Gunther, and Theo van Leeuwen. *Multimodal Discourse: The Modes and Media of Contemporary Communication.* New York: Oxford University Press, 2001.

———. *Reading Images: The Grammar of Visual Design.* London: Routledge, 1996.

Laflin, P. *The Salton Sea: California's Overlooked Treasure.* Indio, CA: The Periscope, Coachella Valley Historical Society, 1995.

Lake, Copeland. "Huge Trailer Hauls Penstock Pipe at Boulder Dam." In *The Story of Hoover Dam,* 129–30 (Las Vegas: Nevada Publications, 1986).

Layton, Edwin T. "Technology as Knowledge." *Technology & Culture* 15 (1974): 31–41.

Leach, Neil, ed. *Rethinking Architecture: Reader in Cultural Theory.* New York: Routledge, 1997.

Leslie, Jacques. *Deep Water: The Epic Struggle over Dams, Displaced People, and the Environment.* New York: Farrar, Straus and Giroux, 2005.

Limerick, Patricia Nelson. *The Legacy of Conquest: The Unbroken Past of the American West.* New York: W. W. Norton, 1987.

———. *Something in the Soil: Legacies and Reckonings in the New West.* New ed. New York: W. W. Norton, 2001.

Lipset, Seymour Martin. "The Work Ethic, Then and Now." *Journal of Labor Research* 13 (1992): 45–54.

Longinus. *On Great Writing (On the Sublime).* Translated by G. M. A. Grube. New ed. Indianapolis, IN: Hackett, 1991.

Lozowick, Louis. *William Gropper.* East Brunswick, NJ: Associated University Presses, 1983.

Lucaites, John Louis, and Robert Hariman. "Visual Rhetoric, Photojournalism, and Democratic Public Culture." *Rhetoric Review* 20 (2001): 37–42.

Lyotard, Jean-François. "The Sublime and the Avant-Garde." In The Inhuman: Reflections on Time, translated by Geoffrey Bennington and Rachel Bowlby, 89–107. Stanford: Stanford University Press, 1991.

MacCannell, Dean. *The Tourist: A New Theory of the Leisure Class.* Berkeley: University of California Press, 1999.

MacEachren, Alan M. *How Maps Work: Representation, Visualization, and Design.* New York: Guilford Press, 1995.

Macmullen, Margaret. "Love's Old Sweetish Song." *Harper's,* October 1947, 371–80.

Marchand, Roland. *Advertising the American Dream: Making Way for Modernity, 1920–1940.* Berkeley: University of California Press, 1986.

Marr, David. *Vision: A Computational Investigation into the Human Representation and Processing of Visual Information.* Cambridge, MA: MIT Press, 2010.

Marx, Leo. *The Machine in the Garden: Technology and the Pastoral Ideal in America.* 35th ed. Oxford: Oxford University Press, 1964.

McBride, Dennis. "Cliff Segerblom (1915–1990)." *Nevada Historical Society Quarterly* 33 (1990).

McCracken, Robert D. *Las Vegas: The Great American Playground.* Reno: University of Nevada Press, 1997.

McGreevy, Patrick Vincent. *Imagining Niagara: The Meaning and Making of Niagara Falls.* Amherst: University of Massachusetts Press, 1994.

McKinsey, Elizabeth R. *Niagara Falls: Icon of the American Sublime.* Cambridge: Cambridge University Press, 1985.

McKnight, Natalie. "Dickens, Niagara Falls and the Watery Sublime." *Dickens Quarterly,* June 2009.

McQuarrie, Edward F., and David Glen Mick. "Visual Rhetoric in Advertising: Text-Interpretive, Experimental, and Reader-Response Analyses." *Journal of Consumer Research* 26 (1999): 37–54.

Mead, Elwood. "Construction of Boulder Dam." *Literary Digest* 116 (1933): 15.

Melosh, Barbara. "Manly Work: Public Art and Masculinity in Depression America." In *Gender and American History Since 1890,* edited by Barbara Melosh, 155–81. London: Routledge, 1993.

Menon, Elizabeth K. "The Functional Print in Commercial Culture: Henry Somm's Women in the Marketplace." *Nineteenth Century Art Worldwide* 4 (2005).

Merchant, Carolyn. *The Death of Nature: Women, Ecology, and the Scientific Revolution.* New York: Harper and Row, 1980.

———. *Earthcare: Women and the Environment.* New York: Routledge, 1996.

———. *Reinventing Eden: The Fate of Nature in Western Culture.* New York: Routledge, 2003.

Messaris, Paul. *Visual Literacy: Image, Mind, and Reality.* Boulder, CO: Westview Press, 1994.

———. *Visual Persuasion: The Role of Images in Advertising.* London: Sage, 1997.

———. "What's Visual About 'Visual Rhetoric'?" *Quarterly Journal of Speech* 95 (2009): 210–23.

Miall, David S. "Foregrounding and the Sublime: Shelley in Chamonix." *Language and Literature* 16 (2007): 155–68.

Mihesuah, Devon Abbott. *Natives and Academics: Researching and Writing About American Indians.* Lincoln: University of Nebraska Press, 1998.

Miller, Perry. *The Life of the Mind in America: From the Revolution to the Civil War.* New York: Mariner Books, 1970.

Mirzoeff, Nicholas. *An Introduction to Visual Culture.* London: Routledge, 1999.

———. "On Visuality." *Journal of Visual Culture* 5 (2006): 53–79.

———, ed. *The Visual Culture Reader.* 2nd ed. New York: Routledge, 2002.

Mitchell, Timothy. "Caspar David Friedrich's Der Watzmann: German Romantic Landscape Painting and Historical Geology." *Art Bulletin* 66 (1984): 452–64.

Mitchell, W. J. T. "Holy Landscape: Israel, Palestine, and the American Wilderness." *Critical Inquiry* 26 (2000): 193–223.

———. *Iconology: Image, Text, Ideology.* Chicago: University of Chicago Press, 1986.

———. *Landscape and Power.* 2nd ed. Chicago: University of Chicago Press, 1994.

———. *Picture Theory: Essays on Verbal and Visual Representation.* New ed. Chicago: University of Chicago Press, 1994.

———. "Showing Seeing: A Critique of Visual Culture." *Journal of Visual Culture* 1 (August 2002): 165–82.

———. "There Are No Visual Media." *Journal of Visual Culture* 4 (2005): 257–66.

———. *What Do Pictures Want? The Lives and Loves of Images.* New ed. Chicago: University of Chicago Press, 2006.

Moeller, Beverley B. *Phil Swing and Boulder Dam.* Berkeley: University of California Press, 1971.

Morton, Timothy. *Ecology Without Nature: Rethinking Environmental Aesthetics.* Cambridge, MA: Harvard University Press, 2007.

Moxey, Keith. "Visual Studies and the Iconic Turn." *Journal of Visual Culture* 7 (2008): 131–46.

Mucha, Jiri. *Alphonse Mucha: Master of Art Nouveau.* New York: Tudor, 1967.

Muir, John. *The Yosemite.* New York: Century, 1921.

Mulholland, Catherine. *William Mulholland and the Rise of Los Angeles.* Berkeley: University of California Press, 2002.

Napier, John. *Hands.* Revised by Russell H. Tuttle. Princeton, NJ: Princeton University Press, 1993.

Nash, Roderick Frazier. *The Rights of Nature: A History of Environmental Ethics.* Madison: University of Wisconsin Press, 1989.

———. *Wilderness and the American Mind.* 4th ed. New Haven, CT: Yale University Press, 2001.

Natanson, Nicholas. *The Black Image in the New Deal: The Politics of FSA Photography.* Knoxville: University of Tennessee Press, 1992.

Neal, Arthur G. *National Trauma and Collective Memory: Major Events in the American Century.* Armonk, NY: M. E. Sharp, 1998.

Newhall, Beaumont. *The History of Photography: From 1839 to the Present.* 5th ed. New York: Bulfinch, 1982.

Nicolson, Marjorie Hope. "The Sublime in External Nature." In *Dictionary of the History of Ideas.* New York: Charles Scribner's Sons, 1974.

Novak, Barbara. *American Painting of the Nineteenth Century: Realism, Idealism, and the American Experience.* 3rd ed. New York: Oxford University Press, 1980.

Nye, David E. *America as Second Creation: Technology and Narratives of New Beginnings.* Cambridge, MA: MIT Press, 2004.

———. *American Technological Sublime.* Cambridge, MA: MIT Press, 1994.

O'Connor, Francis V. *Art for the Millions: Essays from the 1930s by Artists and Administrators of the WPA Federal Art Project.* Greenwich, CT: New York Graphic Society, 1973.

O'Gorman, Ned. "Longinus's Sublime Rhetoric, or How Rhetoric Came into Its Own." *Rhetoric Society Quarterly* 34 (2004): 71–89.

Oravec, Christine L. "The Sublime." In *Encyclopedia of Rhetoric,* edited by Thomas O. Sloane, 757–60. New York: Oxford University Press, 2001.

Park, Marlene, and Gerald E. Markowitz. *Democratic Vistas: Post Offices and Public Art in the New Deal.* Philadelphia: Temple University Press, 1984.

Pettitt, George. *So Boulder Dam Was Built.* Berkeley, CA: Lederer, Street, and Zeus, 1935.

Philippon, Daniel J. *Conserving Words: How American Nature Writers Shaped the Environmental Movement.* Athens: University of Georgia Press, 2004.

———. *The Truth of Ecology: Nature, Culture, and Literature in America.* New York: Oxford University Press, 2003.

Pisani, Donald J. *To Reclaim a Divided West: Water, Law, and Public Policy, 1848–1902.* Albuquerque: University of New Mexico Press, 1992.

———. *Water, Land, and Law in the West: The Limits of Public Policy, 1850–1920.* Lawrence: University Press of Kansas, 1996.

———. *Water and American Government: The Reclamation Bureau, National Water Policy, and the West, 1902–1935.* Berkeley: University of California Press, 2002.

Plochmann, George Kimball. "Plato, Visual Perception, and Art." *Journal of Aesthetics and Art Criticism* 35 (1976): 189–200.

Pohl, Frances K. *Framing America: A Social History of American Art.* London: Thames and Hudson, 2002.

Powers, Alan. "Art Deco: An Image Problem?" *Apollo* 142 (1995): 3–6.

Propen, Amy. "Visual Communication and the Map: How Maps as Visual Objects Convey Meaning in Specific Contexts." *Technical Communication Quarterly* 16 (2007): 233–54.

Przyblyski, Jeannene M. "History Is Photography: The Afterimage of Walter Benjamin." *Afterimage* 26 (1998): 8.

Pursell, Carroll W., Jr. "Government and Technology in the Great Depression." *Technology & Culture* 20 (1979): 162–74.

Rauchway, Eric. *The Great Depression and the New Deal: A Very Short Introduction.* New York: Oxford University Press, 2008.

Reid, John B., and Ronald M. James. *Uncovering Nevada's Past: A Primary Source History of the Silver State.* Reno: University of Nevada Press, 2004.

Reisner, Marc. *Cadillac Desert: The American West and Its Disappearing Water.* Revised ed. London: Penguin, 1993.

Rentmeister, Cäcilia. "Berufsverbot Für Die Musen. Warum Sind so Viele Allegorien Weiblich?" *Ästhetik Und Kommunikation* 25 (1976): 92–112.

Ricoeur, Paul. *From Text to Action.* Evanston, IL: Northwestern University Press, 1991.

———. *Lectures on Ideology and Utopia.* Edited by George H. Taylor. New York: Columbia University Press, 1986.

Ritivoi, Andreea Deciu. *Paul Ricoeur: Tradition and Innovation in Rhetorical Theory.* Albany: SUNY Press, 2006.

Robinson, Michael C. *Water for the West: The Bureau of Reclamation, 1902–1977.* Chicago: Public Works Historical Society, 1979.

Rocha, Guy Louis. "The IWW and the Boulder Canyon Project: The Final Death Throes of American Syndicalism." *Nevada Historical Society Quarterly* 21 (1978): 2–24.

Rodgers, Daniel T. *The Work Ethic in Industrial America, 1850–1920.* Chicago: University of Chicago Press, 1978.

Rosaldo, Michelle Zimbalist. "Women, Culture, and Society: A Theoretical Overview." In *Woman, Culture, and Society,* edited by Michelle Rosaldo and Louise Lamphere, 17–42. Stanford: Stanford University Press, 1974.

Rosen, Eliot A. "Roosevelt and the Brains Trust: An Historiographical Overview." *Political Science Quarterly* 87 (1972): 531–57.

Rosenberg, Daniel. "No One Is Buried in Hoover Dam: Modernism and Rumor."

In *Modernism Inc.: Body, Memorial, Capital,* edited by Jani Scandura and Michael Thurston, 84-106. New York: New York University Press, 2001.

Rothman, Hal K. "Conceptualizing the Real: Environmental History and American Studies." *American Quarterly* 54 (September 1, 2002): 485-97.

———. *Devil's Bargains: Tourism in the Twentieth-Century American West.* Lawrence: University Press of Kansas, 1998.

Rusco, Elmer R. *Good Time Coming? Black Nevadans in the Nineteenth Century.* Santa Barbara, CA: Greenwood Press, 1976.

Russell, R. Robert. "Highlights on Boulder Dam." *Rainbow,* December 1935, 4-7.

Russo, Mary. *The Female Grotesque: Risk, Excess and Modernity.* New York: Routledge, 1994.

———. "Female Grotesques: Carnival and Theory." In *Writing on the Body,* edited by Katie Conboy, Nadia Medina, and Sarah Stanbury. New York: Columbia University Press, 1997.

Scattergood, E. F. "Engineering and Economic Features of the Boulder Dam." *Annals of the American Academy of Political and Social Science* 135 (1928): 115-22.

Scharff, Virginia J., ed. *Seeing Nature Through Gender.* Lawrence: University Press of Kansas, 2003.

Scott, James C. *Seeing Like a State: How Certain Schemes to Improve the Human Condition Have Failed.* New Haven, CT: Yale University Press, 1998.

Scott, Linda M. "Images in Advertising: The Need for a Theory of Visual Rhetoric." *Journal of Consumer Research* 21 (1994): 252-73.

Sears, James M. "Black Americans and the New Deal." *History Teacher* 10 (1976): 89-105.

Sears, John F. *Sacred Places: American Tourist Attractions in the Nineteenth Century.* New ed. Amherst: University of Massachusetts Press, 1999.

Seavey, Clyde L. "What the Boulder Dam Project Means to California and to the Nation." *Annals of the American Academy of Political and Social Science* 135 (1928): 127-32.

Shaffer, Marguerite S. *See America First: Tourism and National Identity, 1880-1940.* Washington, DC: Smithsonian, 2001.

Shaw, Philip. *The Sublime.* New York: Routledge, 2005.

Simich, Jerry L., and Thomas C. Wright. *The Peoples of Las Vegas: One City, Many Faces.* Reno: University of Nevada Press, 2005.

Skerrett, R. G. "America's Wonder River—the Colorado." In *The Story of Hoover Dam,* 11-16. Las Vegas: Nevada Publications, 1986.

Slaton, Amy E. *Reinforced Concrete and the Modernization of American Building, 1900-1930.* Baltimore: Johns Hopkins University Press, 2001.

Slichter, Gertrude Almy. "Franklin D. Roosevelt and the Farm Problem, 1929-1932." *Mississippi Valley Historical Review* 43 (1956): 238-58.

Smith, Henry Nash. *Virgin Land: The American West as Symbol and Myth.* New ed. Cambridge, MA: Harvard University Press, 1978.

Smith, Kenneth L. "Perception and the Newspaper Page: A Critical Analysis." In

Handbook of Visual Communication: Theory, Methods, and Media, edited by Kenneth L. Smith, Sandra Moriarty, Gretchen Barbatsis, and Keith Kenney, 81–98. Mahwah, NJ: Lawrence Erlbaum, 2004.

Sontag, Susan. *On Photography.* New York: Picador, 1977.

———. *Regarding the Pain of Others.* New York: Picador, 2004.

Starr, Kevin. *Material Dreams: Southern California Through the 1920s.* New York: Oxford University Press, 1990.

Steinberg, Theodore. "Fertilizing the Tree of Knowledge: Environmental History Comes of Age." *Journal of Interdisciplinary History* 35 (2004): 265–77.

———. "'That World's Fair Feeling': Control of Water in 20th-Century America." *Technology & Culture* 34 (1993): 401–9.

Stevens, Joseph E. *Hoover Dam: An American Adventure.* Norman: University of Oklahoma Press, 1990.

Stewart, Susan. *On Longing: Narratives of the Miniature, the Gigantic, the Souvenir, the Collection.* Durham, NC: Duke University Press, 1993.

———. *The Open Studio: Essays on Art and Aesthetics.* Chicago: University of Chicago Press, 2005.

Stockton, Sharon. "Engineering Power: Hoover, Rand, Pound, and the Heroic Architect." *American Literature* 72 (2000): 813–41.

Stormer, Nathan. "Addressing the Sublime: Space, Mass Representation, and the Unpresentable." *Critical Studies in Mass Communication* 21 (2004): 212–40.

Stott, William. *Documentary Expression and Thirties America.* Chicago: University of Chicago Press, 1986.

Stovall, Tyler. "National Identity and Shifting Imperial Frontiers: Whiteness and the Exclusion of Colonial Labor After World War I." *Representations* 84 (2003): 52–72.

Striner, Richard. "Art Deco: Polemics and Synthesis." *Winterthur Portfolio* 25 (1990): 21–34.

Sturken, Marita, and Lisa Cartwright. *Practices of Looking: An Introduction to Visual Culture.* 2nd ed. New York: Oxford University Press, 2009.

Sundstrom, William. "Down or Out? Unemployment and Occupational Shifts of Urban Black Men During the Great Depression." *Research in Economic History* 16 (1996): 127–55.

Tagg, John. *The Burden of Representation: Essays on Photographies and Histories.* Minneapolis: University of Minnesota Press, 1993.

Taylor, Charles. "Modern Social Imaginaries." *Public Culture* 14 (2002): 91–124.

———. *Modern Social Imaginaries.* 3rd ed. Durham, NC: Duke University Press, 2004.

Taylor, Ken. "Culture or Nature: Dilemmas of Interpretation." *Tourism, Culture & Communication* 2 (1999): 69–84.

Thompson, Jan. "The Role of Women in the Iconography of Art Nouveau." *Art Journal* 31 (1971): 158–67.

Tichi, Cecelia. *Shifting Gears: Technology, Literature, Culture in Modernist America.* Chapel Hill: University of North Carolina Press, 1987.

Trachtenberg, Alan. *Classic Essays on Photography.* Stony Creek, CT: Leete's Island Books, 1981.

————. *Reading American Photographs: Images as History, Mathew Brady to Walker Evans.* New York: Hill and Wang, 1989.

Trelease, Frank J. "Arizona V. California: Allocation of Water Resources to People, States, and Nation." *Supreme Court Review* (1963): 158–205.

Turner, Frederick Jackson. *The Significance of the Frontier in American History.* New ed. New York: Dover, 1996.

Urry, John. *The Tourist Gaze: Leisure and Travel in Contemporary Societies.* 2nd ed. London: Sage, 2002.

Valocchi, Steve. "The Racial Basis of Capitalism and the State, and the Impact of the New Deal on African Americans." *Social Problems* 41 (1994): 347–62.

Van de Wetering, Sarah Bates, David H. Getches, Lawrence MacDonnell, and Charles F. Wilkinson. *Searching Out the Headwaters: Change and Rediscovery in Western Water Policy.* Boulder: Natural Resources Law Center, University of Colorado School of Law, 1993.

Van Horn, Carl E., Herbert A. Schaffner, and Ray Marshall, eds. *Work in America: An Encyclopedia of History, Policy, and Society.* Illustrated ed. Oxford: ABC-CLIO, 2003.

Vilander, Barbara, Ann. *Hoover Dam: Photographs of Ben Glaha.* Tucson: University of Arizona Press, 1999.

————. "The Hoover Dam Photographs of Ben Glaha Completed Under the Auspices of the United States Bureau of Reclamation." PhD diss., University of California at Santa Barbara, 1995.

Wagner, Rob Leicester. *Red Ink, White Lies: The Rise and Fall of Los Angeles Newspapers, 1920–1962.* Upland, CA: Dragonflyer Press, 2000.

Warren, Karen J., ed. *Ecofeminism: Women, Culture, Nature.* Bloomington: Indiana University Press, 1997.

Wehr, Kevin. *America's Fight over Water: The Environmental and Political Effects of Large-Scale Water Systems.* New York: Routledge, 2004.

Weinberg, Johnathan. "I Want Muscle: Male Desire and the Image of the Worker in American Art of the 1930s." In *The Social and the Real: Political Art of the 1930s in the Western Hemisphere,* edited by Alejandro Anreus, Diana L. Linden, and Jonathan Weinberg, 115–34. University Park: Pennsylvania State University Press, 2006.

Weymouth, F. E. "Conservation of the Waters of the Colorado River from the Standpoint of the Reclamation Service." *Science* 56 (July 21, 1922): 59–66.

Whipple, John, and Estevan López. "New Mexico's Experience with Interstate Water Agreements." *New Mexico Water: Past, Present, and Future or Guns, Lawyers, and Money,* October 2005.

White, Richard. *"It's Your Misfortune and None of My Own": A New History of the American West.* Norman: University of Oklahoma Press, 1993.

————. *The Organic Machine: The Remaking of the Columbia River.* New York: Hill and Wang, 1996.

————. *The Roots of Dependency: Subsistence, Environment, and Social Change Among the Choctaws, Pawnees, and Navajos.* Lincoln: University of Nebraska Press, 1988.

White, Theophilus Ballou. "Building the Big Dam." *Harper's,* June 1935, 113–21.

Wilkinson, Charles F. *Crossing the Next Meridian: Land, Water, and the Future of the West.* Washington, DC: Island Press, 1993.

Williamson, Judith. *Decoding Advertisements.* London: Marion Boyars, 1978.

Wilson, Edmund. "Hoover Dam." *New Republic,* September 2, 1931, 66–69.

Wilson, Richard Guy. "American Modernism in the West: Hoover Dam." In *Images of an American Land: Vernacular Architecture in the Western United States,* edited by Thomas Carter, 291–319. Albuquerque: University of New Mexico Press, 1997.

———. "Machine-Age Iconography in the American West: The Design of Hoover Dam." *Pacific Historical Review* 54 (1985): 463–93.

———. "Massive Deco Monument." *Architecture* 72 (December 1983): 45–47.

Wilson, Rob. *American Sublime: Genealogy of a Poetic Genre.* Madison: University of Wisconsin Press, 1991.

Wilton, Andrew, and Tim Barringer. *American Sublime: Landscape Painting in the United States, 1820–1880.* Princeton, NJ: Princeton University Press, 2003.

Wolf, Donald E. *Big Dams and Other Dreams: The Six Companies Story.* Norman: University of Oklahoma Press, 1996.

Worster, Donald. *Dust Bowl: The Southern Plains in the 1930s.* New York: Oxford University Press, 1979.

———. "Hoover Dam: A Study in Domination." In *Dams: Studies in the History of Civil Engineering,* edited by Donald C. Jackson. Surrey, UK: Ashgate, 1998.

———. *Nature's Economy: A History of Ecological Ideas.* 2nd ed. Cambridge: Cambridge University Press, 1994.

———. *A River Running West: The Life of John Wesley Powell.* New York: Oxford University Press, 2001.

———. *Rivers of Empire: Water, Aridity, and the Growth of the American West.* New York: Oxford University Press, 1985.

———. *Under Western Skies: Nature and History in the American West.* New York: Oxford University Press, 1992.

Worster, Donald, and Constance B. Schulz. *Bust to Boom: Documentary Photographs of Kansas, 1936–1949.* Lawrence: University Press of Kansas, 1996.

Wright, Harold Bell. *The Winning of Barbara Worth.* Chicago: Book Supply Company, 1911.

Wrobel, David M., Patrick T. Long, and Earl Pomeroy, eds. *Seeing and Being Seen: Tourism in the American West.* Lawrence: University Press of Kansas, 2001.

Young, Robert. *Colonial Desire.* London: Routledge, 1995.

Index